CARBON QUEEN

CARBON QUEEN

The Remarkable Life of Nanoscience
Pioneer Mildred Dresselhaus

Maia Weinstock

The MIT Press
Cambridge, Massachusetts
London, England

The MIT Press would like to thank the anonymous peer reviewers who provided comments on drafts of this book. The generous work of academic experts is essential for establishing the authority and quality of our publications. We acknowledge with gratitude the contributions of these otherwise uncredited readers.

This book was set in ITC Stone Serif Std and Pangram by New Best-set Typesetters Ltd. Printed and bound in the United States of America.

Library of Congress Cataloging-in-Publication Data

Names: Weinstock, Maia, author.
Title: Carbon queen : the remarkable life of nanoscience pioneer
    Mildred Dresselhaus / Maia Weinstock.
Description: Cambridge, Massachusetts : The MIT Press, [2022] |
    Includes bibliographical references and index.
Identifiers: LCCN 2021005672 | ISBN 9780262046435 (hardcover)
Subjects: LCSH: Dresselhaus, M. S. | Carbon. | Nanotubes. |
    Physicists—Biography.
Classification: LCC TA455.C3 W45 2022 | DDC 620.1/93—dc23
LC record available at https://lccn.loc.gov/2021005672

10 9 8 7 6 5 4 3 2 1

For X, the most marvelous collection of atoms in the universe

When raising kids was women's work
when homemaking was an aspiration

she debunked tradition, swept others up in her contrails
flew in the face of inequality
donned the epaulettes of a doctorate
destabilizing the norms
in academic male humdrum
in the atom-splitting world of science

her brilliance spawned theories and theorems
unlocked the secrets of carbon's electronic structure
the mysterious forms it takes in nature
predicting the existence of carbon nanotubes—
those single-atom-thick cylinders of carbon for use in
stronger materials, ultra-strong cables, hydrogen storage,
advanced electronics, solar cells, batteries

Working alongside her husband
who never left her in his considerable shadow,
she gave birth to
four children
and pioneering work on
vibrational mechanical energy

And now, well into her 80s, with myriad awards
added to the epaulettes, when the trail
for women has been long blazed
I wonder whether she realizes that her footprint
matches one of the first on the path

—Carolyne Van Der Meer, "The Queen of Carbon" (2016)

CONTENTS

AUTHOR'S NOTE

This book provides a history of the life of physicist and engineer Mildred S. Dresselhaus. In it, I also pay considerable attention to her husband and longtime professional collaborator, Gene Dresselhaus, and to other family members with the same or similar last names. Going back to her youth, Mildred Dresselhaus was widely addressed by her nickname, Millie. While intending to convey both respect and clarity, this book often employs first names alone for Millie and Gene Dresselhaus and for other members of their extended family.

PROLOGUE

Miss Sophia Harvey, seven going on eight, stands before her dining room table, transfixed by a powder-blue bag. She is surrounded by a coterie of family and friends who have gathered to celebrate her eight years with cupcakes, presents, and games. The gift bag seems just the right size, and Sophia can barely contain her excitement at the possibility that someone may have gotten her the one thing she's been wishing for all year. Like a grizzly fishing for salmon, Sophia plunges her hand into the bag, grasps firmly at the object within, and yanks out a navy cardboard box. Inside, a Barbie-style action figure in the likeness of an octogenarian woman, gray hair pinned up in braids, stares back at her. "A Millie Dresselhaus doll!" Sophia gasps. "This is the best birthday ever!"

On a soundstage in New Delhi, Sunita Rajawat makes final preparations for her guest, a personal hero she has anticipated interviewing for months. Thanks to a stroke of magic, the producers of her show—one of the most popular in India—have managed to book the interview on Millie Dresselhaus Day (#MillieDay, on social media), an unofficial holiday celebrated annually on November 11, the eminent scholar's birthday. After a brief meeting offstage, the time has come to introduce her guest to the studio audience, and to the millions of

viewers watching remotely. As the audience provides a standing ovation, Sunita feels her pulse rise. She takes Millie's hand,
welcomes her to the stage, and proclaims enthusiastically, "We
are so glad to have you here!"

It's midday in Toronto, and Millie Dresselhaus is enjoying a
leisurely lunch with longtime colleague Olivier Rhéaume at
their favorite seafood joint. On this brilliant afternoon, she
and Olivier sip cool drinks and talk nanotubes over Faroese
herring and locally caught trout. The paparazzi are out in
force, snapping away as Millie nibbles on her fish. She man-
ages to lose them after the meal, when the restaurateurs allow
her to slip out the back. But later that afternoon, a pair of
young men run her down on the sidewalk: "Une photo avec
vous?" they ask, breathlessly. She agrees, and one of the men
snaps a selfie. After thanking her profusely and wishing her a
pleasant stay in Canada, he posts the memento to Instagram
with the inscription, "Regardez qui est . . . MILLIE #MILLIE
#REINEDECARBONE." By nightfall, the post has over 722,000
likes, and counting.

The preceding fictional vignettes, starring the real-life scientist
and engineer Mildred "Millie" Dresselhaus, were presented in
a 2017 advertisement from General Electric as a vision of what
the world might be like if prominent women in science were
treated as A-list celebrities. (I've added names and story details
to the vignettes to help bring the video to life here on these
pages.) The spot was meant to at once celebrate one of the
most accomplished researchers of our time, inspire the next
generation of women in STEM (science, technology, engineering, and math), and inform the public about the company's
plans to hire more women in technical roles. In addition
to being the model for a doll and a target of paparazzi, Millie became the namesake of numerous newborns; the most

popular dress-up costume at Halloween; and an icon whose image was plastered on cityscapes, T-shirts, and teenagers' text messages. The video was a resounding success and, in real life, went viral on social media.[1]

Yet this brief star turn for Millie Dresselhaus barely skimmed the surface of what she accomplished in her remarkable eighty-six years. Millie's research forever altered our understanding of matter—the physical stuff of the universe that has mass and takes up space. A longtime professor of electrical engineering and physics at the Massachusetts Institute of Technology, Millie also played a significant role in inspiring people to use this new knowledge to solve some of the world's greatest challenges, from producing clean energy to curing cancer.

As her nickname suggests, the Queen of Carbon is most often heralded for her pioneering work with one of the world's most abundant and versatile substances. Most of us know carbon as the smooth, slippery graphite of a pencil tip; the alluring shimmer of a diamond ring; and the Jekyll and Hyde of energy—both the basis of hydrocarbons that we transform into electricity and petroleum-based commodities, as well as the by-product of fossil fuel combustion that's warming Earth's atmosphere at an alarming rate. Carbon is also the stuff of life, the element that makes compounds organic and one of the few elements present in all of the living organisms we know about. Scientists say there are about 10 million carbon compounds, with more being discovered each day.[2]

As a result of Millie Dresselhaus's insatiable curiosity about our world and her nearly six-decade career as a scientific explorer, we can thank her for significant leaps in how we think about carbon's various forms and the company it keeps. In her early career, Millie employed a recent invention—laser light—to probe carbon's inner workings, to distinguish how, for example, flat layers of carbon atoms act differently from carbon crystals of three dimensions, especially in the presence

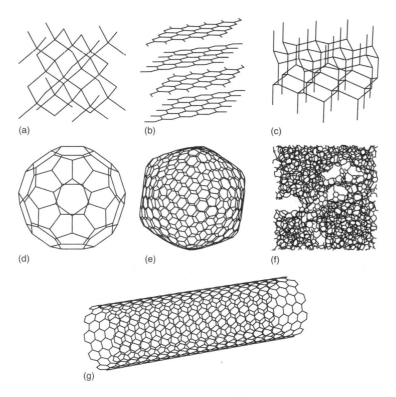

FIGURE 0.1

Over nearly sixty years, Mildred "Millie" Dresselhaus uncovered some of carbon's fundamental properties—especially relating to its electronic structure. Here we see seven carbon configurations, or allotropes: (a) diamond; (b) graphite; (c) lonsdaleite; (d, e) fullerenes C_{60}, also known as a buckminsterfullerene, and C_{540}; (f) amorphous carbon; and (h) a single-walled carbon nanotube.

of heat, electrons, or a magnetic field. And she predicted the existence of what we now call carbon nanotubes, sheets of carbon atoms rolled up into minuscule cylinders that can be remarkably adept at conducting electricity. Based on Millie's far-reaching foundational research, scientists and engineers have made enormous advances at the nanoscale—with structures on the order of one-hundred-thousandth the width of a human hair. Spherical carbon "buckyballs," cylindrical carbon nanotubes, and two-dimensional carbon sheets known as graphene have been developed for use in energy storage, medical research, building materials, and paper-thin electronics, among many other applications. In the past ten years alone, the pace of discovery has skyrocketed. These carbon structures continue to be developed for myriad novel uses, with quite a few seemingly out of the realm of science fiction, from ultrafast quantum computers, to efficient desalination devices, to electronics-infused bodily organs.[3]

In an era in which, for the first time in history, a significant portion of all human knowledge can be called up, Star Trek–like, at any time from a tiny screen on your wrist, it may seem difficult to imagine that computer science hadn't even come into its own as an academic discipline when Millie Dresselhaus began her career in 1958. It was only four years earlier, in 1954, that silicon was first used to fabricate transistors, semiconducting devices that allow the controlled flow of electrons, and came to serve as the basis for modern computing machines. Today's engineers are just beginning to exploit the electronic properties of carbon, in a wave of technological progress that may one day lead to the development of a Carbon Corridor to match Silicon Valley. And while numerous scientists and engineers have helped pave the way toward a future of carbon-based technologies, Millie Dresselhaus pioneered both theories and experiments that led to a fundamental shift in what we know about the electronic structure of carbon and the fascinating things that happen when researchers stir bits of other elements into carbon samples.[4]

Nanoscientists in particular owe a massive debt to work that Millie Dresselhaus carried out during her multifaceted career. At its simplest, you can think of nanoscience as the study of material structures between 1 and 100 nanometers in size. (A nanometer is a billionth of a meter.) As a whole, nanoscience is an extraordinarily diverse line of inquiry, with major contributions from disciplines as varied as biology, chemistry, computer science, electrical engineering, materials science and engineering, mechanical engineering, and physics. The field has helped to uncover the secrets behind a host of materials, but carbon is inarguably one of the most intriguing from the perspective of potential nanoscale applications. As a result, it's safe to say the Queen of Carbon played an outsized role as a founding pioneer of nanoscience and consequent nanotechnologies, all of which continue to grow in leaps and bounds.

In all, Millie authored or coauthored an astounding 1,700 research articles and eight books, largely relating to carbon and its fundamental properties. She spoke the languages of science and engineering as if they were her mother tongue. But she was far more than a brilliant researcher. She was also a tireless educator and role model for both the young people she worked with and for those near and far who aimed to follow in her footsteps. Her mentorship activities evolved as she transitioned from young professor—one of the first women on the MIT faculty—to internationally recognized advocate for science and engineering, but throughout her career, she was a beloved figure, providing countless individuals with support at every stage along their journeys in academia or elsewhere.

"Millie didn't just create a lab, she created a family. She worked with us to ensure we succeeded," wrote Sandra Brown, former student and now associate general counsel at Rensselaer Polytechnic Institute, in a 2017 tribute. "She was also a truly genuine and sincere person. One of the proudest days of my life was when Millie met my parents, blue-collar workers and first-generation immigrants from a developing country. Millie shook their hands with the kind of warmth she's shown to presidents, princes, and dignitaries. She had a way of making you feel like you mattered, regardless of your status or start."[5]

Beyond her mentorship and advocacy, Millie was also a trailblazer whom countless women in science and engineering looked up to. She was the first female MIT Institute Professor, the highest distinction for faculty at MIT. She was the first woman to win a National Medal of Science in the category of engineering. And she was the first solo recipient of the prestigious Kavli Prize, given biennially in the disciplines of astrophysics, nanoscience, and neuroscience. To be sure, such firsts are often little more than reflections of the awarding institutions. But firsts beget notability, and notability begets influence, and when your line of work continues to suffer from significant underrepresentation of certain groups, being a first can have considerable tangible effects, by inspiring, for example, individuals who might not otherwise consider venturing into engineering or physics or mathematics—or who might simply need a little support.

Millie knew this and used her influence as a pioneer whenever she could. She did so as a leader of numerous national societies representing scientists and engineers around the United States. She did so for innumerable international colleagues. And she did so at her home institution of MIT—often quietly and without fanfare.

From her earliest years in Depression-era New York City, Millie embodied resilience. She could have made a fortune in lemonade, given how she handled the many lemons that came hurtling toward her and her family. Fending off challenging environments as well as active discouragement from

certain educators and advisers in her youth and early adult-
hood, she never let anyone derail her curiosity—and she made
every effort to pay it forward for others.

While she may not have been a media celebrity with a massive global following as in the ad that celebrated her life, Millie
was nothing less than a hero to many, in her home country
and around the world. The story of her life—which began
in immense poverty, involved a childhood battle against a
deadly disease, and required persistence through low expectations and sexist attitudes to become a highly influential and
revered scientist—provides insights into the workings of an
ever-expanding mind, the ongoing evolution of societal attitudes toward women and people of color in science and engineering, and the unique power of kindness.

1 DIAMOND IN THE ROUGH

In the shadow of the Brooklyn Bridge, where the storied span meets the bustling Brooklyn-Queens Expressway, a quiet, unassuming stub of Sands Street hides its secrets well. For half a century, this once-thriving thoroughfare was punctuated on its western end by a massive transit hub that connected Brooklyn to the rest of New York City and pulsed as the nerve center of the borough's sprawling elevated rail system. The three-story terminal also served as a gateway to the nearby Brooklyn Navy Yard, the "Can-Do Shipyard" that produced warships in the nineteenth and twentieth centuries, from the USS *Ohio* to the USS *Constellation* (CV-64), employing some seventy thousand men and women during its heyday in the 1940s.[1]

Today, the byway connecting these two landmarks, some ten minutes by foot, still exists at the nexus between several Brooklyn neighborhoods. But any semblance of its more colorful, pedestrian-friendly past—of shops, bars, or the transient navy personnel it served—is long gone. In place of the former Sands Street rapid transit station, which closed in 1944 and was later demolished, a series of access ramps now cuts through the western end of the street, allowing some ten thousand pedestrians and cyclists and over sixty thousand motor vehicles to climb toward the bridge and across the East River into Manhattan every day.[2]

It is here, at 45 Sands, across from what is now a parking lot, that Mildred "Millie" Dresselhaus, one of the most influential scientists in modern history, began her remarkable life. Shortly after her birth in Brooklyn's Flatbush neighborhood, Millie's immigrant parents, Meyer and Ethel Spiewak, moved here as the new owners of a newsstand and candy shop. Mere steps from the looming Sands Street terminal, which would inhale Brooklyn's working class during the morning rush and spit them back out after eight-, ten-, and twelve-hour shifts, the shop offered magazines, fresh fruit, and bottles of chilled Coca Cola among other refreshments. It was a meager living, to be sure, but one that afforded Meyer and Ethel some stability and a means to pay their $25-a-month rent.[3]

Millie's parents had come to America to escape an increasingly unstable Europe in the years after World War I. Her father, Meyer Spiewak, hailed from the Polish village of Dzialoszyce, where the family trade had been singing; his father, brother, and grandfather had all been synagogue cantors. (*Spiewak* means "singer" in Polish.) Millie's mother, Ethel Teichtheil, was originally from Galicia, a former Austro-Hungarian region that today encompasses parts of southeastern Poland and western Ukraine. Ethel's mother had died suddenly when Ethel was a child, and after the outbreak of World War I, her father moved the family to Holland for safety. There, Ethel became well educated for a girl at the time; she had at least eight years of formal schooling and excelled in languages. Like his granddaughter would be, Ethel's father was an expert in carbon—namely, diamonds, in which he traded. The Spiewak family also worked in the diamond business, and Millie's parents met and became a couple as their families worked closely together.[4]

But as tensions between nations grew even deeper in Europe following the end of World War I, Ethel and Meyer became anxious about their future. After considerable deliberation, they decided they would be better off in America and made plans to begin new lives on the other side of the Atlantic. Meyer arrived in New York in 1921, and the couple married two days after Ethel joined him, on December 13, 1926. It was a prescient move: as Millie would recount in later years, nearly everyone in the Spiewak and Teichtheil families who stayed in Europe perished at the hands of the Nazis during the Holocaust.[5]

Though guaranteed certain basic freedoms in the United States, the transition to life in New York City was not an easy one for the Spiewak family, which grew in 1928 with the birth of Millie's older brother, Irving, and again on November 11, 1930—Armistice Day (now Veterans Day)—when baby Mildred made the family four.[6] The Sands Street area they moved to, which today includes sections of Brooklyn Heights, downtown Brooklyn, DUMBO, and the Navy Yard, was rough. In his biography of the infamous gangster Al Capone—who, as a child, lived south of the Navy Yard's main entrance at Sands Street's eastern end—author Robert J. Schoenberg wrote, "Sands offered all the diversions suggested by the phrase 'drunken sailors.' In uniforms customized at Max Cohn's, sailors could swagger into the saloons that lined Sands, then stagger out to the attendant pickup dance halls, brothels, hot-bed rooming houses, pawn shops, and tattoo parlors."[7] Indeed, Millie would later relay to her own children a story for the ages: of the night when she was quite small, no more than about three years old, when a drunken sailor broke into her family's apartment, making a rowdy, fearsome scene and scaring everyone witless. This part of Brooklyn, it turned out, was also home to dangerous gangs, and the Sands Street station, with few police present, was often teeming with thieves.[8]

Making matters eminently worse was the economic black cloud that came storming through the United States a year before Millie was born. Lasting through the 1930s, the Great

Depression colored much of Millie's childhood and early life, in Brooklyn, and especially so in the Bronx, where Millie spent her formative years. It would, among other things, make holding a job difficult for her father and force her mother to work long hours to put food on the table. It also compelled Millie to work odd jobs—some of which skirted child labor laws—on top of her schoolwork to help the family make ends meet.[9]

And yet a powerful calling would soon creep into Millie's world, one that would forever alter her trajectory. It was a pursuit that, as a young girl with limited resources, enabled her to see what in the world was possible. As an adult, it helped her to develop a repertoire with some of the most accomplished academics in the world, both when she was making her way and when she eventually became a highly regarded scholar and mentor herself. Throughout her life, this calling filled Millie's home and her spirit, providing her with a distinctive language with which to communicate—one she mastered even before learning to read. The calling was music, and thanks to Millie's older brother, it became her primary avocation for the rest of her long and storied life.[10]

BROOKLYN TO THE BRONX

For young Irving Spiewak, playing the violin was like breathing: making his tiny instrument sing required no determined action or thought; he moved his fingers, and the notes just came. His talent was not an overwhelming surprise to his family, as generations of Spiewak men before him had been skilled musicians. But by the time he was in preschool, Irv's prowess on the violin made others take notice. When one violin teacher in particular, one of the finest in New York City, met five-year-old Irving, he felt strongly that the child had the makings of a prodigy and requested taking him on as a scholarship student.[11]

Irving's parents, eager for their son to take advantage of every opportunity for success—especially if it meant developing a musical talent—agreed to the arrangement. Being a scholarship student, Irv would apprentice with his teacher, lessons completely gratis, several times a week. The only glitch was that Irving's teacher was based in the Bronx, at the Bronx House Music School, a daunting commute from Sands Street in Brooklyn. Irv and Millie's parents came up with a plan: if they sold their bodega and moved to the Bronx, Irv would be much closer to his teacher, and they wouldn't have to spend so much time transporting him to lessons. It was a calculated risk; it would cost the Spiewaks just about all of their resources to relocate, and Meyer would need to find a new job to support the family. In their minds, however, the gamble was worth it. Music was part of the family's blood, and Irving's promise seemed legitimate. They buckled down and rolled the dice.[12]

As was often the case for American immigrants settling in New York in the early 1930s, things did not quite turn out as planned. Just months after the Spiewaks gave up all they had to move into a meager tenement apartment in the Bronx, Irving's violin teacher died suddenly. The loss was an enormous blow, and it set off a chain of events that would keep the family in poverty for quite some time. And yet, as with several other pivotal moments in Millie's life, the setback came with silver linings that would reveal themselves in time, leading ultimately to key opportunities for both Irving and his kid sister.[13]

Having depleted their assets in relocating to the South Bronx, at that time a poor community with a large population of Central and Eastern European immigrants, the family had no means to pay for a new violin teacher.[14] The United States was already deep into the Depression, with newspaper headlines reading like SOS alerts from doomed vessels: "Industrial Depression—Unemployment—Destitution! Idleness and Want

Drive Men to Crime and Suicide! Desperate Situation Demands
Serious Consideration!"[15] Meanwhile, Meyer Spiewak, who
had become a day laborer after giving up the candy shop, was
having trouble finding steady work to provide for his wife and
children. Despite the meager assistance they received in the
form of government-sponsored welfare, following Irving's talent was something the Spiewaks felt they had to do; if nothing
else, his gift might prove to be a source of much-needed cash
in the not-too-distant future. Another teacher had to be found
who would take him on pro bono.[16]

Thanks to Irving's tremendous abilities, a replacement was
identified in short order, but the individual was once again
a long commute away, this time at the Greenwich House
Music School on the opposite side of the city, in downtown
Manhattan—exactly the kind of situation Meyer and Ethel
had hoped to avoid by moving to the Bronx. Bound to accept
a teacher who would offer services free of charge, there was
little choice in the matter: The family had pinned its future on
Irving's star, and they couldn't move again, so the apprentice-
ship was arranged.[17]

It was around this time that young Millie Spiewak began
showing musical promise of her own. It wasn't the violin
that impressed the adults around her; that would come later.
Rather, Millie had developed a gift for musical memory. As
a four-year-old tagalong at her brother's lessons, she demon-
strated that she could remember the melodies of many compositions and sing them flawlessly. This seemed to the teachers at
Irving's music school to be the sign of another prodigy in the
making. And so, Mildred joined her older brother in obtaining
a scholarship to study music herself in the heart of Greenwich
Village. Little did she know how far-reaching this endeavor
would become, shaping the course of her life and career in
a multitude of ways.[18] "Music school had a big influence on
me," she noted in a 1976 oral history recorded at MIT, her

longtime academic home. "As a matter of fact, without music school, I never would be where I am today—not in any way."[19]

A HARD-KNOCK LIFE

Millie's admittance to the Greenwich House Music School coincided with her introduction to the neighborhood she would call home for nearly twenty years. Today, all that remains of her former walkup at 1631 Washington Avenue in the Bronx is a nondescript parking lot. On a given day you might find it, three blocks south of the teeming Cross-Bronx Expressway, on the northwest corner of East 172nd Street, littered with trailers, a smattering of parked cars, and piles of empty shipping pallets. The street outside is lined with New York City Fire Department ambulances, standing at attention, while a medical services station and a recently modernized firehouse sit adjacent to the lot. Across the street sits an immense—and immensely drab—industrial park housing a plant run by one of the largest pharmaceutical manufacturers in the world. If you buy prescription or over-the-counter drugs at outlets like CVS or Walmart, there's a decent chance they were made here.[20]

Back in the 1930s, however, 1631 Washington Avenue was the site of a cookie-cutter tenement house, home to scores of huddled masses who'd landed in a poverty-stricken neighborhood where crime was rampant. When Millie's family arrived here, the Great Depression was in full effect, but it wasn't the only social force shaping everyday life for New Yorkers. The tail end of Prohibition, the nationwide alcohol ban that ended in 1933, had served for thirteen years as a petri dish for bootleggers and gangs—mostly recent immigrants of Irish, Italian, Jewish, and Polish descent who smuggled in liquor and sold it on the black market.[21] "I stayed home whenever I could," Millie admitted in her 1976 MIT oral history. "It was kind of dangerous in the streets, and my parents were always very much fearful about having us out of sight."[22]

Her parents' concern was warranted. As Millie would later recall, she had been on the receiving end of random assaults by other children walking to and from school and other local destinations. "Oh, people were beaten up all the time," she explained. "I was attacked by gangs of kids a few times and bruised here and there; [I] sometimes came home a bloody mess."[23]

The violence didn't end in the streets. When Millie began taking classes at her local public elementary school, she found the other students simply could not focus on schoolwork—and they had a tendency to be wild, acting out in class and in the schoolyard. Her teachers could barely keep students in line and, as a result, would be able to squeeze in real lessons only once in a blue moon. By junior high, things hadn't improved.[24] "The teachers had almost no time for teaching," Millie said in a 2015 interview. "They would just try to keep order, which wasn't an educational experience. . . . We were told to go to the bathroom at home before we went to school, so we wouldn't have to go while we were there. It seemed that going to the bathroom was a bit dangerous because girls would get mugged."[25]

Millie's parents instituted strict rules about going outside unchaperoned. They generally discouraged Irving and Millie from participating in sports and other physical activities. If they wanted to play with other children, another adult had to be present or they weren't allowed to participate. In keeping with family tradition, however, Meyer and Ethel strongly encouraged music and education for their children, so unaccompanied subway rides were allowed during Millie's elementary and junior high school years, as long as the destination was an educational activity. This meant that Millie, by around age six, would traipse through the city to her music school by herself.[26] "Going from the elevated train to the subway involved many steps," she noted in a 2012 interview. "I remember how, many times, I fell down those steps with my violin and my books."[27]

In the 1930s and early 1940s, Millie's life was constrained in another important way: she had to work. At a time when the national unemployment rate had jumped into the 20s, Millie's father often had a hard time finding a job. Making matters more difficult, he was frequently ill, which further limited his ability to bring home a paycheck. This compelled Millie's mother, Ethel, to look for employment outside the home. She managed to find a twelve-hour-a-day overnight job caring for children at an orphanage, which helped somewhat but left her exhausted. The job paid meager wages, so despite her enormous efforts, the family continued to rely on public assistance to make ends meet. It was a grueling existence, and Ethel's perseverance in making sure the family's basic needs were met would influence Millie throughout her life.[28] "Some days we didn't have anything to eat," Millie recalled in a 2001 interview. "But I do remember when it was really bad we had potato soup. I still remember potato soup because potatoes were pretty cheap, and you could boil up potatoes in a big kettle of water, and it would make a lot of fluid."[29]

On top of their food insecurity, the Spiewaks could afford very little in the way of material possessions, and young Millie had no toys. "My mother was very, very careful that I was absolutely clean and neat, my hair was combed," Millie stated in 2001, "but I had only one set of clothes and I wore [them] every day."[30]

Millie's first paying job began in elementary school, when she was eight years old. She was paid fifty cents a week for fifteen to twenty hours of work tutoring a special-needs student who had severe difficulty retaining information.[31] "It wasn't very good pay, but the experience was fantastic because this child was a super challenge," Millie said in 2001. "I had to entice him somehow to study, and I learned a lot; I worked out to be a good teacher with him."[32]

By this time, Millie's own teachers could sense she was well beyond her peers in terms of curiosity and drive. When she reached junior high, she was told flat-out by one teacher that attending class would be a waste of her time; instead, the teacher suggested that Millie become an administrative helper—an activity she later said provided her with early organizational skills and a glimpse at how to be a manager.[33]

The future physicist also had chores at home, as well as odd jobs to supplement the family income. With her mother taking on the role of chief breadwinner, Millie was expected to pick up domestic duties that her brother never had to concern himself with: cleaning, meal preparation, and the like.[34] According to one of her granddaughters, Shoshi Dresselhaus-Cooper, "Millie said she was never resentful about this, and never thought it was unfair. . . . It was only in reflection that Millie decided her added responsibilities [were] probably a better lesson than her brother ever got, because doing all that when she was a kid taught her how to get an impossible amount of work done!"[35]

In addition to her unpaid work, for a few hours each day Millie also helped with manufacturing assembly work that her mother would bring home to supplement the family income. "I was always good with my hands at making things," she noted in 1976. "We used to make jewelry and piecework. You get paid by how many of the different things you made per hour."[36]

And then there was the zipper factory. To help the family stay afloat, Millie worked as a child laborer in a New York City sweatshop putting zippers together during her summers off from school.[37] "That was pretty hard work," she later admitted. "Every single operation that had to be made on them, I did at one time or another while I was working in that factory."[38]

Memories of this time came flooding back to Millie in the 2000s during a family outing to catch the 1936 Charlie Chaplin classic *Modern Times*, a comedy in which Chaplin plays a factory worker struggling to adjust to the realities of the Great Depression and the twentieth-century rise of industrialization and manufacturing.[39] After emerging from the theater, Millie reflected on her long-gone zipper days, noting to her family that she used to hide when inspectors came around because she was under the legal minimum age for workers in New York City.[40]

DOWNTOWN SANCTUARY

When she could steal away to the Greenwich House Music School, Millie was able to let her guard down, forget the weight of the world, and explore life as a wide-eyed child. During her first years there, a bright, young Millie blossomed into a veritable sponge, soaking up every experience and opportunity that crossed her path.[41]

Compared with her troubled South Bronx neighborhood, Millie's beloved Greenwich Village hangout could not have been a more different environment—and it served as something of a safe house for her and her brother. To Millie, each subway ride into Greenwich Village was like a jaunt in Cinderella's carriage, whisking her away from the tiresome reality of daily life into a fantasyland where her every curiosity was piqued, and instead of working labor-intensive tasks to help her teachers and her parents, she could explore her musical, artistic, and intellectual leanings.[42]

While she had signed up to be there for music lessons, the community at Greenwich House—one of a series of cooperative cultural centers around New York City that still function today—provided much more than just arpeggios and the circle of fifths. "We want to take the child out of his humdrum

environment, expose him to facets he didn't know existed,"
explained Maxwell Powers, a Greenwich House conductor in
a 1950 interview with the *New York Times*.[43] Millie was one
scholarship student who benefited tremendously from this
outlook, for it was at Greenwich House that her peers and
their families provided her with a vision for how she might
rise above the challenging circumstances of her youth.[44]

Although Millie had a tremendous amount of talent and
could read musical notation before reading cursive English,
she turned out not to be the prodigy her brother was. To be
sure, she took numerous violin lessons in her many years with
the program. But she admitted later that she had lacked both
the passion and motivation required to become a virtuoso.
Still, Millie loved being at Greenwich House, and the education she received and people she met would stay with her
throughout her life.[45]

In exchange for her scholarship, the Greenwich House
administrators occasionally asked Millie to run errands for
them. These she would do with pleasure, in part because she
and the school's other scholarship students were preferentially
offered free tickets to see concerts and theatrical performances.
Ever the go-getter, Millie wasted no time in taking advantage
of as many of these opportunities as she could. In this way,
she was able, for example, to catch an early Leonard Bernstein
performance, well before the Lawrence, Massachusetts, native
became a world-renowned composer and conductor.[46]

But Millie's greatest perk—aside from her free violin lessons—was the position she earned as a film critic for the
Greenwich House Music School newsletter. Paying her way
into theaters to see big-name films, Millie's instructors introduced her to more exciting new worlds and provided opportunities for the young aesthete to travel the globe without
leaving Manhattan.[47]

Disney's *Fantasia*, which arrived in theaters exactly two days after Millie's tenth birthday, was one dramatic work that made, in her words, "a terrific impression."[48] Beyond the well-known Sorcerer's Apprentice, the delightful tale of curiosity and magic set to music by Paul Dukas and starring the lovable and timeless Mickey Mouse, more abstract segments, such as the animated Toccata and Fugue in D Minor, based on the composition by Johann Sebastian Bach, gave symphonic music—its scales and vibrations and repeated patterns—visual life in a way that no other work had done before. The meandering melody of a flute, for example, could be seen emerging from the din of supporting instruments as a singular contrail, drifting aimlessly through a billowing fog of fuchsia and crimson clouds, while dramatic refrains from the strings section punctuated the screen as beams of light that ultimately joined together in a white curtain of sound.[49]

Millie saw the film multiple times, appreciating it even more after studying some of the featured compositions at her music school. But beyond the theatrics that transported her to another time and place, Millie also benefited from the writing assignment attached to each film she experienced. According to granddaughter Shoshi Dresselhaus-Cooper, Millie really learned to write when she crafted movie reviews for the Greenwich House newsletter. She also gained valuable lessons from the careful and detailed critiques that her supervisors would return to her as comments on her reviews—something she herself was exalted for with her own students and colleagues later on.[50]

Millie's most treasured episode during her tenure at Greenwich House involved a special guest who inspired her beyond words and would remain a personal hero for the rest of her days. This individual came around to visit the school on several occasions thanks to her friendship with Greenwich House's inspirational founder, social worker and city planner Mary Kingsbury Simkhovitch. The visitor was Eleanor Roosevelt.[51]

In her syndicated newspaper column, "My Day," Roosevelt noted her connection with Greenwich House on at least two occasions. "This evening," she wrote in July 1939, "we are going to preview the movie 'They Shall Have Music,' given for the benefit of the Greenwich House Music School and the High School of Music and Art. I shall tell you about it tomorrow."[52]

The drama in question follows Frankie, a musically gifted runaway who discovers a New York City settlement house providing free music lessons for the poor. To raise money in order to keep the house afloat, Frankie convinces famed real-life violinist Jascha Heifetz to perform a benefit for the cause.[53] "I enjoyed it very much," Roosevelt wrote of the film the following day, "and hope that many people will see it and be led to support music schools for poor children of talent, for that is the purpose to which this picture is dedicated."[54]

While Roosevelt was a patron of Greenwich House, Mary Kingsbury Simkhovitch was an unapologetic supporter of progressive causes, including the political career of Roosevelt's husband, President Franklin D. Roosevelt. As a result of their overlapping interests, Eleanor Roosevelt took the opportunity to visit Greenwich House every now and again to take in concerts and other happenings. On one of these occasions, young Millie Spiewak was asked to play for the first lady—an event that would be seared in her mind for the rest of her life.[55] "That was really quite remarkable," Millie said in a 2016 family interview. "It was kind of amazing to have such an important person—the first lady of our land—coming to our student recitals and encouraging the kids. But that was Mrs. Roosevelt. That was her personality."[56]

The encounter encapsulated more than just a young admirer playing for an idol. It also provided key inspirations that Millie would return to again and again: that someone as important and internationally known as the first lady would take the time to visit with city children of diverse backgrounds and that Roosevelt had boldly stepped out of earlier molds for president's spouses and pursued independent and progressive activities of her own on behalf of the American people—especially those who were disadvantaged.

SCIENCE INSPIRATIONS

Back in the Bronx, with all of the work that was expected of her, and with her music lessons in Manhattan—and long subway rides to and fro—young Millie had little time to play with other kids. As a result, she didn't have many close friends during her early childhood. But she managed to find time for activities that gave her joy—and that hinted at her exceptional future.[57]

She was fascinated, for example, by Paul de Kruif's 1926 classic *Microbe Hunters*, a collection of essays lionizing fourteen men of science (and their colleagues and confidants), from Antonie van Leeuwenhoek to Paul Ehrlich, all conquerors in humanity's war against pestilence and disease. This best-selling work of narrative nonfiction was written with a dramatic flair that both humanized scientists and rendered them heroes—a perfect way to enthrall a bright, curious youngster.[58]

Millie's interest in science was also stoked by an award-winning biography of renowned physicist and chemist Marie Curie. Written by Curie's daughter Eve, *Madame Curie* contained the two-time Nobelist's detailed personal history, including excerpts from numerous letters and correspondences, as well as insights from an author who knew the subject intimately. While it would be many years before Millie could envision a career for herself in such a profession, this book provided an early glimpse of a woman exceling in science.[59]

Additional fascination with *National Geographic* magazine, which Millie could buy with the pocket change she diligently saved up from her tiny allowance, further immersed her young mind in scientific and humanistic thinking. "I really learned science from that because you could get three old copies for 10 cents, and that was about the level of my allowance," she once said.[60]

But despite the injection of science and nature into her life through various activities and resources, Millie's day-to-day routine remained particularly stressful through junior high school. Being poor during the Great Depression took a significant toll; the daily slog of household chores, the charade of attending public school, and her various paid jobs were a continuous test of her mettle. Millie would later admit that getting through her youth was the single biggest hurdle that life had thrown at her. "My greatest challenge was surviving as a child," she noted in a 2004 interview. "That was the hardest thing I ever did."[61]

In typical Millie fashion, however, she came to see the bright side of her strained situation and learned to use it to her benefit. "Tough conditions as a child either make you or break you," she added in 2004. "If you can survive hard times, you have an advantage . . . you have a level of maturity that other people don't have, to overcome adversity."[62]

From an early age, Millie learned to help herself when it came to confronting difficult circumstances. When her neighborhood was dangerously overrun with gang activity and her elementary school experience was one big exercise in futility, she went out of her way to befriend local gang members of every stripe, persuasion, and allegiance. This helped her throughout her time in the Bronx: "The gangs used to fight each other, but if you weren't a member of any gang and you were on good terms with everybody, they would let you pass without attack; I was living in that sort of status when I was a kid, so I suffered relatively little combat compared to a lot of the other kids I knew."[63]

As a youngster, Millie also made deliberate attempts to sow racial harmony in her neighborhood. It was a time when the South Bronx was home to an increasingly diverse mix of people, including émigrés from Europe and the West Indies as well as Black Americans from the contiguous United States—the epitome of the American melting pot. Millie became active in building community bridges between predominantly white neighbors and neighbors of color. "I was one of the kid leaders setting up an interracial settlement house for both Blacks and whites," she explained in 1976.[64] Such efforts and outward friendliness made her a "Millie from the block" of sorts: While she maintained a fairly small circle of close friends, she was respected by everyone she met. This earned her a level of protection that allowed her to roam the streets mostly unperturbed throughout her youth.[65]

TICKET TO A NEW WORLD

If her neighborhood provided Millie with street smarts and a desire do what she could to improve daily life for others, it was her music school that was most responsible for her early intellectual growth. By the time she was in junior high, interest in furthering her education had been firmly drawn, thanks to the people Millie met at Greenwich House. Like her parents, Millie's Greenwich House instructors, as well as her peers and their families, valued education highly. But unlike her parents, who were simply not aware of avenues for advancing Millie and Irving on a nonexistent budget, the individuals she met at Greenwich House provided her with concrete ideas about furthering her educational horizons. She learned, in particular, about New York City's excellent public magnet high schools, which drew the city's best and brightest students.[66]

Millie's brother, Irving, made his way into one of these magnet schools. By the mid-1940s, he had shown prowess not only as a master violinist but as a student of science and mathematics. While he easily could have turned his musical talent into a full-time occupation—and he did ultimately make some money playing various gigs and continued recreationally for the rest of his life—Irv became passionate about scientific inquiry and began to follow a path toward science as a career. So precocious that he skipped ahead several grades, he attended the elite Bronx High School of Science and was well on his way toward graduating early, enrolling in the tuition-free Manhattan engineering school Cooper Union, and then heading up to MIT for a master's in chemical engineering—all by the age of eighteen. In fact, his star was so bright that Millie purposely avoided walking in his footsteps too closely, for fear she could never live up to his example.[67]

By the end of her junior high years, three of New York City's four top magnet high schools—Bronx Science, Stuyvesant, and Brooklyn Tech—still prohibited girls from matriculating. Of these, Bronx Science reversed course the fastest; it welcomed girls for the first time in 1946, just before Millie finished high school. It would be another two decades before Stuyvesant and Brooklyn Tech followed suit—Stuyvesant did so only after being sued for discrimination in 1969 and Brooklyn Tech finally admitted girls for the first time in 1970.[68]

One magnet school in the early 1940s, however, specifically educated girls: Hunter College High School on the Upper East Side of Manhattan. As a public school, any New York City resident could attend free of charge—so long as they passed the rigorous entrance exam. However, Millie's junior high school in the Bronx actively discouraged her from applying. "The teacher said, 'Oh, you have no chance,'" Millie later recalled.[69]

Considering that getting into Hunter was actually more of a feat than getting into one of the other city magnet schools—only one was open to girls, and it admitted fewer than one hundred students a year, while there were three boys' schools, each accepting more students per class—her teachers weren't entirely wrong to be dubious about her chances. But the barrier placed in front of Millie only served as a springboard for her already well-developed sense of self-motivation. Once she knew there was a way out of her impoverished childhood, there was no stopping her from attempting to get there.[70] "I didn't have any other option," she later noted. "I knew that to get to the next step, I would have to be a superstar."[71]

Millie thus resolved to conquer the Hunter High admissions test, despite not knowing almost anything on it. "I prepared for the entrance exam all by myself, and kind of kept to myself that I was doing it," she said.[72] She figured out that she could obtain some information directly from the school, including examples of previous entrance exams. "I couldn't even understand the language on these exams; it was like another world. But, New York has very good libraries. . . . I checked out books and got to work."[73]

If this book were a movie, here would be the scene where the enterprising young go-getter struggles by herself to master some slightly insane goal—to the "Chariots of Fire" theme song or similar. She checks out heavy textbooks; she furrows her brow and scratches her head at the indecipherable gibberish in which she must become fluent; she undergoes many fits and starts. But slowly she begins to get it. Finally, on the day of the exam, our protagonist is ready. She knows everything on the test, forward and backward. And she aces it—with nothing less than a perfect score.[74]

2 BRAINS PLUS FUN

Millie plotted her attack in stealth mode. Bored during recess one blustery winter afternoon, she had decided to cause a little trouble: In a well-timed flurry of action, she picked up her first piece of ammunition—a dusty chalkboard eraser—and chucked it like a grenade.

Before her classmates could respond, she quickly maneuvered herself into position to deflect return projectiles, calculated the trajectories needed to cause maximum chalk-dust fallout across the room, and let a second round of erasers fly.

In mere seconds, the room devolved into chaos. Blackboard erasers flew past each other, and the class became ground zero for an epic battle that would have made Braveheart proud. Joining Millie at the center of the eraser fight were a couple of kids who had come from more disadvantaged schools prior to arriving at Hunter High. Millie's other classmates, mostly from well-to-do backgrounds, found the scene so absurd that they quickly joined in by egging the others on.

"Then suddenly the teacher walked in," Millie later recalled, "and she got hit in the face with a board eraser that I threw."[1]

Oops.

The teacher was so flustered that she said absolutely nothing and went right into her lesson. To Millie's recollection, no one was punished. "And that ended that; never more were there board eraser fights at Hunter High," she mused.[2]

Millie Spiewak's acceptance into the prestigious Hunter College High School in Manhattan was a critical step in her journey toward an unlikely career in science. But despite her obvious intellectual gifts and an inner drive to succeed, she would face numerous obstacles along the way, in high school, graduate school, and beyond as she blazed a path that would lead her to the pinnacle of the scientific world.

When Millie matriculated at Hunter High in February 1945, she was one of just a few students in her grade who had attended a local junior high rather than a feeder school in the Hunter College system. While she had become a diligent violin student at the Greenwich House School of Music and advanced as a gifted self-learner in her academic subjects, she maintained her friendships with gang members and other gregarious neighbors and was used to sitting in classrooms where her peers routinely acted out, teachers struggled to maintain order, and little actual instruction was achieved. She was quite independent for her age, and a bit of a sparkplug when it came to seeking out new experiences. It would be fair to say, then, that upon entering prim and proper Hunter High, Millie was something of a fish out of water.[3]

To be sure, she was already well ahead of the curve when it came to her drive to learn, be it in academic realms or more humanistic, worldly experiences and skills. But Millie's penurious background, in terms of scant home resources and the limited opportunities offered by her neighborhood, would help to mold her into a teenager who took matters into her own hands and who made sure to appreciate the joy in life's little pleasures. Since her parents did not encourage Millie to participate in sports or other physical activities, however, she lacked traditional outlets for childhood competition and other age-appropriate high jinks. So it was not entirely uncommon for her to engineer such experiences—a blackboard eraser fight or other sneaky scenarios—for herself.[4]

MILLIE TAKES MANHATTAN

The start of Millie's high school years coincided with the tail end of both the Holocaust and World War II, intertwined upheavals in the course of human events that at once shattered affected families—including Millie's own, which had largely been destroyed by the Nazi regime—and also ushered in a gradual easing of poverty conditions for many in the United States.[5]

Around this time, Millie's mother was able to quit her demanding overnight job at the orphanage and transition into a new position at a leather factory, making wallets. It was a bump financially thanks to a recent increase in the federal minimum wage; the new job also had regular business hours and happened to be fairly close to Millie's school on the Upper East Side, so Ethel and Millie began sharing an hour-long commute together most mornings.[6]

Between this newfound one-on-one time and the fact that they were sometimes the only two at home, mother and daughter came to form a particularly close bond. Millie's brother by this time was no longer a regular part of the household; in his midteens, Irving was already in college studying chemical engineering at Cooper Union in Greenwich Village and rarely returned home. Meanwhile, Millie's father struggled mightily after losing most of his family following the devastation in Europe. He continued to bounce around between jobs and sometimes required medical care; occasionally he spent long spells in a hospital, leaving Millie and Ethel to fend for themselves.[7]

From an early age, Millie had enjoyed a great deal of freedom of movement; in elementary school, her parents had let her spend long hours on the train by herself riding to music lessons in Manhattan, as well as to other educational activities around town. Now that she was commuting long distance to high school, that movement picked up precipitously in support of her studies, her school activities, and her personal activities—both organized and unorganized.[8]

Millie also continued to make new acquaintances, though she kept in touch with a few bosom buddies from her Bronx neighborhood, whom she tended to bring along with her on jaunts around the city. These individuals, whom she called "exceedingly devoted," would come looking for Millie to lead them on new adventures.[9] "Traveling was 5 cents on the subway at the time; it would take you everywhere in the city," Millie said. "I could go anywhere, at any time."[10]

One of Millie's favorite activities was exploring the city's world-renowned museums, especially the American Museum of Natural History, as well as art museums. She claims to have at one point known the entire collection of the natural history museum by heart—something that encouraged her interest in science and mathematics when she began taking these subjects in high school.[11]

As a young teenager, Millie became particularly fascinated by astronomy, and she and her friends often visited the acclaimed Hayden Planetarium, which hosted—as it still does today—sky shows featuring immersive storytelling that described the wonders of the cosmos with music and simulated stargazing. To a bright-eyed youngster fascinated with how the world and universe worked, the whole experience was indelible.[12]

There was just one small problem: the Hayden charged admission, and Millie and her friends had no spending money, so this made it impossible for them to purchase tickets. The detail didn't derail her. "I figured out a way to sneak into the planetarium," she explained with a laugh in her 1976 MIT oral history. "I saw every planetarium show that they had . . . but I, of course, never paid to get in."[13]

In one particular instance, the enterprising Millie was surprised by an especially hawk-eyed guard. "Something didn't quite go right, according to the game plan of how you sneak into the planetarium, and I got caught," Millie said. "I told them that I wasn't trying to do anything wrong as far as their collection was concerned; I didn't have any money, and I was interested in astronomy. They just sent me packing."[14]

But Millie wasn't through. After a couple of weeks away— "you know, a cooling off period"—she turned around and "went right back to my old tricks" of getting in.[15]

Sneaking into museums wasn't her only illicit activity around town. Millie went through a phase of being fascinated by theater, but of course, theater tickets—aside from those she received thanks to her Greenwich House connections—weren't free, and in fact could be quite pricey. Millie's solution: hatching a fail-safe method for sneaking into Broadway shows, whereby she would miss the first third of the show, then slip past the ushers, who tended to let their guard down after the production was well underway "There was one period when I had seen all the plays from low-class to high-class," she said. "I couldn't possibly afford to go to them, but I got to see them."[16]

Later, in college, Millie realized that being a volunteer usher would afford her the same opportunities without needing to sneak around. But as a young student, she loved bringing her friends along for the ride. "Some of them were less interested in the things that we saw than in the thrill of sneaking in," she noted.[17]

SECOND TO NONE

At Hunter High, Millie delighted in the intellectual stimulation that her teachers and classmates offered, and she took advantage of as many extracurricular activities as possible. Throughout her life, she would note again and again that Hunter High was a turning point for her career. There, she was able to explore her curiosities in many areas, especially math and science, by taking what she called an "overload" of courses—bonus electives, in other words.[18] "That was what really became my central focus and led to my lifelong career in physics research and teaching," she noted in a 2009 speech recorded for that year's Hunter College High School commencement ceremonies.[19]

At Hunter, Millie grew enormously on an interpersonal level as well. At lunchtime, students tended to pursue activities in school clubs, and Millie found herself making scores of new friends through these groups and through the orchestra, in which she naturally played violin. "I got into all kinds of activities; I belonged to a different club every day," Millie recalled. "Everything was so different from my neighborhood schools, so I started acting like the kids in the high school."[20]

"I met Mildred when we were fourteen," says Elizabeth "Betty" Stewart, one of Millie's oldest friends. She and Millie were both new to the Hunter system when they arrived in 1945 and became pals. "Millie was always a friendly and warm girl and I always felt very comfortable with her. It wasn't long before I realized she was extremely gifted, even though she never boasted or dwelled on her abilities. She seemed to take all her accomplishments in stride, but I was very impressed. She excelled in every subject, was a serious musician, and could even make some of her own clothes. I never knew anyone with such a wide array of gifts."[21]

Since she had taught herself enough math to ace Hunter's entrance exam, Millie found herself in very good shape with mathematics, and she continued to be a stellar math student throughout her time at the school. "Millie's math skills were legendary," Betty Stewart affirms. "I, on the other hand, was having trouble with a pretty simple math class: geometry. I failed the midterm horribly, and in competitive Hunter that was not acceptable. Millie listened to my woes and volunteered to tutor me and get me through. She asked for nothing in return. She came to my house a time or two and patiently explained the basic concepts. With her help, I passed the final exam with a 93. I never forgot the friendship she showed me."[22]

In addition to her schoolwork, Millie had continued through junior high with various odd jobs for the benefit of her family. But starting in high school, she ratcheted her entrepreneurial game up a notch, developing a well-paying tutoring operation spread mainly through word of mouth.[23]

As with the young special-needs student she had previously tutored, Millie was initially recommended as a tutor for a fellow student who was struggling. The pay was substantial: $5 per hour, or $67 an hour in 2021 dollars.[24] At the time, Millie considered this to be "highway robbery," but her stellar reputation and the line of parents who were willing to shell out that kind of money became so long that by the time she got to college, Millie had earned enough to not only help her parents with bills but also to become financially independent.[25] The subjects on offer included her wheelhouse areas of math and science, but she would teach anything: English, history, Spanish—whatever was in demand. "It was my own little business. . . . I almost always used to get an A in the course for my student. I would figure out what their problems were, why they couldn't get it. . . . I would teach them how to study and how to prepare for an exam."[26]

With every client, Millie's tutoring experience grew more advanced—to the point where she began to envision secondary education as a future career. But just as she was hitting her stride in school, in her activities, and in her extracurriculars, a serious illness threatened her very existence.

Early in her junior year of high school, Millie developed a deep, persistent cough accompanied by a high-pitched "whoop" sound after each breath following a cough. This was whooping cough, a disease that affected nearly 110,000 Americans in 1946, the year Millie contracted it.[27] Whooping cough is a dangerous illness that can be deadly in infants and others with underdeveloped immune systems. The bacterium that causes this highly infectious disease was identified in 1906 by Belgian bacteriologists Jules Bordet and Octave Gengou, and a vaccine was developed in the early 1940s thanks to work by pediatrician Leila Denmark and others. Research led by American scientists Grace Eldering and Pearl Kendrick, with support from Loney Gordon, led to the vaccine we know today—but it wasn't until 1943 that the American Academy of Pediatrics approved it for routine use in the United States.[28]

Millie almost certainly was not vaccinated, and she became very sick. She was forced to stay home for many months in order to recover—and to be sure she didn't infect her classmates. According to one interview, this was the reason she finally stopped her violin lessons and other activities at Greenwich House Music School. "I got very sick during the middle of my high school career, and I had to stop something," Millie said. "I missed almost a full school year."[29]

When she eventually returned, fully recovered, Millie set her sights toward finishing strong and planning for her next adventure: college. Even thinking about higher education was fairly uncommon for an American woman in the immediate postwar years. In 1940, only about 77,000 women in the United States earned bachelor's degrees, compared to 110,000 earned by men. By 2017, the number of bachelor's women earned had climbed to 1.12 million, eclipsing the 836,000 that men earned.[30]

Hunter College High School's strong academic program served as an undeniable turning point for Millie's academic outlook, but she would later admit that her experience there also highlighted what she *didn't* know in terms of both scholarly pursuits and the opportunities for young women's advancement in the late 1940s. At the time, women who wanted to attend college had limited options, especially if they had few financial resources. And although numerous women had made headway in traditionally male-dominated STEM fields during World War II, many found their jobs eliminated or otherwise given back to men returning from the war effort. As a result, women—especially those who came from disadvantaged backgrounds—were systematically overlooked for placement and advancement in these areas.[31]

"I had low aspirations," she stated in a 2012 interview. "I was told that because I was a girl and also because I had no money, there were three possible careers for me: school teaching, nursing, and secretarial work."[32]

With these three choices laid out before her, Millie naturally gravitated toward teaching, and with her affinity for mathematics, she planned to become a math teacher. She had practically become one during high school already; with her impeccable tutoring skills and keen business acumen, she had established the 1940s equivalent of a successful educational start-up.[33]

But were there other options? At this stage of her life, Millie lacked role models in the STEM fields—professionals who could show her what the path to becoming a scientist, mathematician, or engineer might look like. And while her teachers were very encouraging, her school's guidance counselor didn't even try to place students at colleges that seemed beyond their financial means. What's more, while the science and math curriculum at Hunter High was on an entirely different plane from what was available at her Bronx neighborhood high school, Hunter maintained a focus on liberal arts education while leaving advanced science and math to its brother institutions. As a result, although she took as many math and science electives as possible, Millie missed out on the opportunity to learn these subjects at a level that might prepare her for a career in one of the STEM fields. And yet she found a way to learn the material anyway. When Millie befriended some of her peers from Stuyvesant High School in Lower Manhattan, she inquired whether they might share some of their knowledge, and they agreed.[34]

Still, when it came time to apply for college, Millie balked. It wasn't that she didn't want to go; rather, she knew that Hunter College High School students were automatically awarded a place at neighboring Hunter College, and Millie assumed she could never go anywhere else for lack of funds—and the need to continue helping her family with the money she was earning from tutoring. She even garnered a prestigious scholarship from Cornell University for her strong achievement on her New York State Regents exams, but she doubted she could afford the balance of tuition and fees, so she didn't apply—there or anywhere else. "To me, it seemed pretty obvious that going to college meant Hunter College."[35]

While it might seem like a lost opportunity to gain an Ivy League education, Millie made out just fine— and in fact most likely wouldn't have turned into the remarkable scientist and engineer that she became were it not for her experience at Hunter College, and for one particular professor who would change her life forever.

On February 3, 1948, Millie floated across the stage at Hunter College High School, brimming with hope for the future. She was a top graduate in her class and the winner of two special awards: the Parent-Teachers' Association Scholarship, "for scholarship and promise of service to the community," and the Hunter College Chapter of Pi Mu Epsilon Award, "for proficiency in mathematics."[36]

Her senior yearbook, the simply titled *Annals*, January 1948 edition, is covered in a futuristic, almost holographic, dark gray woven material with a repeating pattern of crescents that shimmers when you twist the cover under a light. It's almost as if the yearbook's 1940s glam look presaged Millie's future as the discoverer of some of the most fundamental properties of another dark gray material that, when woven into tennis racquets and cell phone covers, appears to shimmer when you twist it just so.[37]

Inside, Millie is listed by yearbook staff as "intelligent and kind." Meanwhile, the description by her portrait is one for the ages.

> MILDRED SPIEWAK: Any equation she can solve; Every problem she can resolve. Mildred equals brains plus fun; In math and science she's second to none.[38]

Sixty-one and a half years later, Millie would address graduates of the same institution, with words of encouragement like she received before heading off to college: "Very few people in the world enjoy their work as much as I do, and I strongly recommend it for all of you. So think big and think broadly. . . . You can do anything that you set your mind to do."[39]

It was no small feat for Millie to achieve all that she did in science and engineering, especially considering how many barriers she and other career-oriented women faced in the mid-twentieth century. But before she could triumph against any of those challenges, she first needed to envision how such a future might materialize. That vision began to solidify just over a year after her high school graduation, thanks to one of the most important mentors of her life.

3 TO TEACH OR NOT TO TEACH

The late 1940s encompassed a unique period for physics in America. On the heels of World War II, researchers were left to reckon with the effects of the Manhattan Project, which provided all-too-real evidence of the previously inconceivable amount of energy packed inside a minuscule atom. Some continued on in nuclear physics, working to harness that immense power for good. Others, like physicist Leo Szilard, shifted their focus; Szilard had conceived of the nuclear chain reaction but was outspoken against its use for atomic weapons and moved into molecular biology after the war.[1]

A decade earlier, Austrian physicist Lise Meitner had made an enormous contribution to nuclear science when she codiscovered fission with Otto Hahn (who went on to win a Nobel Prize while Meitner was left massively snubbed). Meanwhile, scores of women contributed to research within the Manhattan Project, from Leona Woods Marshall Libby and future Nobel laureate Maria Goeppert Mayer to First Lady of Physics Chien-Shiung Wu. Yet in the postwar period, American women were routinely discouraged from pursuing careers in science and engineering. Many top colleges and universities wouldn't even admit women as students on a regular basis until the late 1960s or early 1970s. Women of color were particularly hard to find in labs and in scientific journals during the mid-twentieth century.[2]

This was the climate in which Mildred Spiewak found herself when she began at Hunter College in February 1948. As Millie herself noted in interviews and speeches, her prospects for a career in science were all but nonexistent at that time. Not only was she discouraged by her high school guidance counselor, despite her obvious interest and aptitude, to pursue a college degree in a science- or math-related field, Millie was treated like the many bright, impoverished students—especially those who were women—before her had been: she was not considered a serious candidate for any kind of research career because of her financial standing.[3]

And so Millie began her collegiate experience with low expectations. Despite her childhood inspirations, from *Microbe Hunters* and *Madame Curie* to all of the visits to the Hayden Planetarium, academic research was not something she thought about for long-term pursuits. "I was interested from the beginning in math and science," she explained. "As I studied it, I thought it was a real neat thing to do, but I didn't think about it for a career."[4]

In reality, Millie's top priority was simply to improve on the financial situation her parents had struggled through. "My objective . . . was getting some training, so I could do something better than work in a zipper factory," she noted.[5]

In addition to her inability to visualize applying her math and science skills elsewhere, Millie reasoned that her goal of self-sufficiency had already been met thanks to her lucrative tutoring business, which she continued well into college. Of course, even if she'd ended her college career the way she began it—focusing on elementary mathematics education—she'd have been a smashing success; her tutoring experiences made it clear that Millie could have gone on to be a wonderful school teacher. What actually happened was a little more extraordinary—and far exceeded her and her parents' wildest imagination.[6]

For their part, Millie's parents were pleasantly surprised with her achievements heading into college. While they had encouraged her educational activities from the start, most American girls at the time weren't expected to continue studying beyond high school. So while her parents fully assumed their son would attend college on his way toward a career, Millie's choice to do the same was a bit unexpected. Her parents certainly didn't object to her furthering her education, but they could offer little in the way of guidance.[7]

Millie began college immediately after her high school graduation in winter 1948. Early on, she was excited about earning her education credentials: "I looked forward to training for a career in teaching school children to love learning and science," she noted in 2008.[8] But with advanced standing due to strong high school grades, she was also thrilled to be able to take on a healthy serving of electives, mostly in math and science. At first, these electives were just for fun. Even at Hunter, which generally supported women, educators in the 1940s were not especially supportive of "career women" beyond a few acceptable professions. As a result, someone like Millie, an all-around go-getter and highly gifted student, could have a hard time seeing much of a plan for the future.[9]

As her first term picked up speed, Millie really began to enjoy college life. It wasn't the kind of finding-yourself, first-time-away-from-home existence that most people think of when they picture college students, chiefly because Millie continued to live at home and commuted daily to her classes. But she maintained an active social life and was quite popular.[10]

Then things changed very quickly for her as a sophomore. It was at this time that she met and instantly bonded with someone who would serve as a teacher, a role model, a friend, and even something of a mother figure during their many decades in contact.[11]

AN EMINENT EXEMPLAR

Rosalyn Sussman Yalow is best known as the second woman to win the Nobel Prize in Physiology or Medicine, a feat she accomplished in 1977 for her development of the radioimmunoassay technique, a way to use radioactive labeling to measure concentrations of biological and pharmacological substances in blood. (Yalow shared the Nobel with two others for unrelated work; her longtime collaborator, Solomon Berson, had died and was therefore ineligible for the prize.) The first woman to win the medicine prize, Gerty Cori, had done so exactly three decades earlier, when she shared the 1947 Nobel with her colleague (and husband) Carl Cori and with Argentine researcher Bernardo Houssay for their collective work on sugar metabolism.[12]

For Yalow, the first American-born woman to win any Nobel in science, her path to success was like an ant's course to its nest—meandering but with a singular objective: translating her scientific acumen into a career focused on research. When Yalow met her future protégé, just two years after Cori received her Nobel in Stockholm, she was struggling to find a place for herself within the scientific community. "She'd gotten a PhD and then she couldn't get a suitable research job," Millie noted in a 2002 interview. "She was teaching at Hunter College, where the students were mostly going into education. It was not a place of real research interest."[13]

Yalow in fact had attended Hunter College a decade prior and in the process became the institution's first physics graduate. In an effort to pry open a door to a research career, she worked briefly as a secretary before following an opportunity to teach—and earn a PhD in nuclear physics—at the

University of Illinois. But research positions remained largely closed to women—and Jews—during the mid-1940s, especially after World War II veterans returned from service. She eventually landed a full-time research position at what was then the Bronx Veterans Administration Hospital, where she would remain for the next four decades until her retirement. But prior to finding that research home, Yalow returned to her alma mater as a way to tread water while she figured out her next move. She served as an adjunct professor at Hunter for only a few years, but her timing was incredibly consequential for the trajectory of one student in particular. Were it not for Yalow and her star pupil overlapping for approximately sixteen months at a city college in the country's largest metropolis, the course of Millie's history would have been drastically different.[14]

In February 1949, sophomore Millie Spiewak enrolled in an introductory-level physics course that covered the basics in Yalow's specialty, nuclear physics. For Millie, fascinated by the latest developments in our collective understanding of matter, an opportunity to learn how the human mind could have translated mere curiosity into tools for both improved daily living as well as massive, unyielding destruction was too good to pass up.[15]

When she read the course announcement, she was especially intrigued to learn that the class would be taught by a woman. At Hunter High, Millie had been surrounded by female peers and teachers, but the science faculty there was nowhere near fluent in the latest developments in nuclear physics. Yalow, by contrast, had just finished her doctorate in this area and was as much an expert as anyone teaching such an introductory class.[16]

Millie said the course was "a very exciting" one that "totally knocked me over."[17] With class enrollment in the single digits, student and teacher got to know each other well. They bonded immediately, in what Millie later described as "sort of love at first sight." While Millie found in Yalow a scientist who shared her passion for inquiry and provided strong academic and career encouragement, Yalow saw a bit of herself in the whip-smart Millie, who obviously shared a drive to follow her academic interests, regardless of whatever rules—actual or understood—she had to bend in order to do so.[18]

Yalow became a trusted mentor who would continue to nurture Millie in ways large and small throughout her career. The strongest means of early support was encouraging her to forget teaching and pursue research. "She was an amazing person who in her own right opened the field of biophysics and became a world figure in science," Millie said of Yalow in a 2012 interview. "She was the one who was most influential in leading me to attend graduate school and to go to the best schools and to study with the best scholars. She [told] me that I could make it even though I was a woman, and she did warn me that the road ahead for women in science might be more difficult, but not to be deterred."[19]

To supplement her course work, Yalow suggested that Millie attend colloquia hosted by the Columbia University Department of Physics, home to individuals like Willis Lamb and Polykarp Kusch, who would go on to share a Nobel Prize for work on electrons and hydrogen, and to Chien-Shiung Wu, an expert in radioactive decay whose monumental experiment on the conservation of parity would lead to a Nobel for two of her male colleagues. Yalow also invited Millie to her home on at least one occasion. "That was amazing; no other teacher ever did that," Millie said.[20]

In truth, *encourage* may not be the most accurate word to describe Yalow's early support. According to Millie, once her mentor recognized her talent, she all but insisted that Millie change her plans for the future. "Rosalyn was quite a domineering person," she recalled after winning the prestigious

Kavli Prize in 2012. "She just gave orders, and I pretty much did what she said."[21] In a *New York Times* interview that same year, Millie said of Yalow, "You met her and she said, 'You're going to do this.' She told me I should focus on science. She left the exact science unspecified but said I should do something at the forefront of some area."[22]

The two had different personalities. Millie was generally accommodating, quick to avoid confrontation, and always seeking places where she could quietly make a positive mark, whereas Yalow was singularly headstrong.[23] This could be a positive attribute for someone striving for leadership, especially at a time when women were still seen as inferior to men in science (and many other realms). "She has to be that way," Millie explained to Yalow biographer Eugene Straus in the mid-1990s. "If she weren't that way she wouldn't be what she is today. That very strong focus. The world is gray, but she is able to make black and white out of it, and that's always helped her."[24]

Yet when she took someone under her wing, as she did with Millie, Yalow was extremely loyal. "There are sides of Rosalyn that the public doesn't see but I've seen," Millie noted in a 2002 interview. As an example, Millie recalled that after college but still very early in her career, Yalow would, whenever possible, bring her husband, Aaron, to Millie's brief, ten-minute American Physical Society talks—along with shopping bags brimming with goodies. "She can be very motherly," Millie added.[25]

Millie did take Yalow's exhortations to heart and changed her focus from education to physical sciences. However, she wasn't entirely sure which direction to follow. Although she was fascinated with physics and chemistry, she continued as well with a strong mathematics course load and was seriously considering math as an alternate focus. She also had no real idea of what she'd be doing when she finished her studies. As she continued toward graduate school, she even began getting some pushback from her parents, who thought she might be trying to do too much with her academic proclivities.[26]

Yalow left Hunter to pursue full-time research at the Bronx VA during Millie's junior year, but she remained committed to encouraging Millie to apply for prestigious fellowships in research programs that would lead to graduate degrees. Of course, Millie was pulling her own weight, acing her courses and generally making it difficult for anyone without prejudice to turn her away. She also continued to tutor throughout the year, gaining more and more experience and saving up for the cost of a graduate degree.[27]

Rosalyn Yalow wasn't the only one singing Millie's praises and helping her. Though Hunter was a large public college, its physical sciences departments were tiny, and Millie was a standout student, so her professors in these areas got to know her well. Like Yalow, they did what they could to encourage her to seek out graduate study opportunities.[28]

With lavish praise from Yalow and other Hunter faculty, Millie secured several opportunities for advanced studies as she neared graduation in 1951. One acceptance was at Radcliffe College, where she would study physics by taking classes at "brother" institution Harvard University; this scenario would also allow her to continue dating a young man who was a Harvard graduate student. Another acceptance was a fellowship at MIT. This would extend her path in mathematics and introduce her to computer science with Whirlwind, a project to develop a vacuum tube computer for the US Navy. A third acceptance was a fellowship in the newly established Fulbright Program, which would give Millie the chance to study physics at Cambridge University in the UK.[29]

Millie considered the merits of each program carefully, but the decision that solidified the direction of her career ultimately rested on one factor: travel. For Millie, whose family was so destitute during her childhood that they at times had trouble putting food on the table, the chance to expand her physical horizons by exploring another part of the planet proved impossible to resist. "I was attracted to the idea of going abroad and traveling and seeing the world," Millie later said. "It was more that than my preference for physics over math that made that choice." And that's how the future Queen of Carbon decided her career would lie in physics rather than in mathematics.[30]

Millie also hoped to attend Radcliffe as well. She asked the program managers about the possibility of deferring her admission for a year so she could complete her Fulbright before starting a program in Massachusetts. The administrators agreed. And although she turned down the MIT math fellowship outright, she would eventually find other ways to make her mark at the Institute, which in due time was to become her home away from home for more than half a century.[31]

On the evening of June 21, 1951, nearly a thousand young women and men gathered to mark the successful completion of their degrees earned at Hunter College. Following musical selections from the New York Fire Department band and the Hunter College Choir, the new graduates, decked out in Hunter purple caps and gowns, accepted awards and diplomas for their years of dedicated scholarship. For most in the audience, the occasion marked the last stop in their formal education; for a rarefied few, it was just the beginning.[32]

This was certainly a day to remember for Millie. It had been an influential three and a half years, thanks especially to her new friend and champion, Rosalyn Yalow, who dramatically reframed the possibilities for Millie's future. While she later admitted most of her science and math courses had been limited in their depth and outlook, she had clearly benefited greatly from Hunter's overall philosophy. "I learned a lot of things there, in terms of the responsibility of an individual to society, that it's not enough to only take, but you have to give," she later said.[33]

Millie was one of only five students in her class to graduate summa cum laude—with highest distinction. In her graduation program, she was listed with numerous honors, from membership in Phi Beta Kappa to receipt of two Joseph Gillet Memorial Prizes, one in mathematics and one in physics.[34] But perhaps the most memorable aspect of the day was her interaction with the ceremony's featured speaker: Mina Rees, director of mathematical sciences in the US Office of Naval Research. After the commencement ceremony, Rees congratulated Millie specifically. She "knew about many places and many things," Millie said of the mathematician who would become the first female president of the American Association for the Advancement of Science. Rees discussed with Millie her plans to pursue a Fulbright fellowship in England, passed on her strong approval, and encouraged Millie to continue with her studies. "It was," said Millie of the exchange, which stuck with her for decades afterward, "a nice pat on the back."[35]

TAKING FLIGHT

For twenty-year-old Millie, the prospect of a year at the University of Cambridge in England was like a proto-moth finding out she was about to emerge from her cocoon: While she had been a ravenous, experience-seeking caterpillar during her high school and undergraduate years, being able to fly on her own for the first time was something else altogether.

Millie's Fulbright experience marked the first time that she could examine both technical and cultural interests at her own pace, without financial or other burdens. It also was the first time she could explore a bit of the continent from which her family hailed. "It was sort of a year that was mine," she later said.[36]

Millie's academic base was the highly regarded Cavendish Laboratory, otherwise known as the Cambridge Department of Physics—home to more than a dozen Nobel laureates by the time she arrived (and twenty-nine as of this writing). Lectures were open to anyone who wanted to attend and featured a wide variety of subjects, so the year she spent in Cambridge helped to bolster her basic understanding of both high-energy physics—particle physics, delving into the fundamental nature of atoms—and solid-state physics—the study of large-scale properties of matter.[37]

Because of the limited opportunities for deep engagement in physics at Hunter College, Millie's course work at the Cavendish served in particular to introduce her to subjects and processes that she was not already familiar with. Millie had free rein in terms of the areas she pursued, and she chose a healthy mix of science courses and courses in more humanistic areas. She delighted, for instance, in learning a bit of art history, as she found Cambridge hosted "terrific art lectures."[38] On the physics side, Millie vastly improved her understanding of theory—but there didn't seem to be any space for her in the teaching labs, so her experimental work remained lacking until she began her PhD program.[39]

One individual on the Cambridge faculty would come to play an outsized role in Millie's trajectory: Brian Pippard, a professor. He was well known in the physics community for his work in superconductivity, the seemingly magical ability of certain materials to conduct electricity without any resistance when they are cooled below a particular threshold temperature. Superconductivity is, among other things, a property that makes magnetic resonance imaging machines work for diagnosing disease, keeps maglev (magnetic levitation) trains free flowing, and enables high-energy accelerators, such as the Large Hadron Collider, to smash subatomic particles together. Millie and Pippard crossed paths at Cambridge when Millie attended his lectures, and their relationship would blossom a few years after Millie's fellowship, when Pippard visited the University of Chicago, where Millie was pursuing her doctorate.[40]

As part of the Fulbright experience, Millie was also assigned tutors, known colloquially as dons, who worked with her one-on-one in a kind of independent study program. "It was sort of like going to a music lesson," Millie explained; "they give you a lot of work to do, and you come back with the work done the next week and you talk about it." Millie's dons included Robert Dingle, who specialized in condensed-matter physics; Robert Chambers, who worked in solid-state physics; and Tony Lane, whose expertise was in high-energy nuclear physics.[41]

When she wasn't studying physics or boosting her future prospects as a researcher, Millie maintained a highly active lifestyle. She lived in housing reserved for international students and mingled liberally with her peers—both fellow Fulbright students and English locals. Among her activities, Millie dazzled her new friends and acquaintances with her strong violin skills, and she sang in the King's Chapel Choir.[42] Residential life resembled something out of a magical boarding school fantasy novel: "Tea is a study break . . . and eating your meals is conversation time," Millie recalled. "Meals were served around these big tables, and then once a month or so, you ate at the high table with the dons."[43]

During her fellowship year, Millie was also living comfortably for the first time in her life: the Fulbright Program paid for all of her living expenses, plus travel. In short, for anything she wanted to do, she was taken care of. "I had so much money!" she would recall. "Oh, I lived like a queen in England."[44]

Millie used the travel funds for exploration whenever she could. On weekends, she often accompanied her peers on excursions to their hometowns in municipalities throughout the UK. But she also took advantage of Cambridge's long vacations and traveled extensively to other parts of Europe, which was still reeling from years of conflict. She tried to visit family wherever possible, and in particular made regular trips to see an uncle who lived in Paris. She was especially taken with the history of Europe, which she appreciated as steeped in tradition.[45] "There were so many things that were very new to me and were so different from things in the US," she recalled. "And then one could see how other people lived; that was important."[46]

In the end, Millie's year in Cambridge had an enormous impact on quite a few fronts. She was pleased with the faculty, fellow students, and overall program in terms of preparing her to be a strong candidate for graduate school. "It was very, very alive and exciting" at Cambridge in the early 1950s, she said.[47] The program also forced Millie to teach herself a great deal, which helped to solidify in her mind that she could succeed as a scientist. In this way, Millie's Fulbright permanently altered the course of her life's work. "I learned that I could manage to do science at a satisfactory level for me, and that was a good feeling," she said.[48] She also learned "a lot of professionalism" by meeting physicists and physics students from all walks of life and seeing what it was like to be an academic researcher.[49]

Millie translated her enthusiasm into newfound determination to pursue an advanced degree. To embark on this new journey, she recrossed the Atlantic and settled in Cambridge, Massachusetts.

IN SCIENTIA VERITAS

For the first 243 years of its existence, the oldest institution of higher learning in the United States catered exclusively to male students. Things began to change at Harvard University in

1879—the same year a boy named Albert Einstein first graced
the world with his presence—when a new opportunity opened
up for women who wanted (and had the means) to further
their education. That year, with encouragement from banker
and philanthropist Arthur Gilman, a group of seven women
led by natural historian Elizabeth Cary Agassiz established
what became known as "The Harvard Annex," a program in
which members of the Harvard faculty would be paid on top
of their regular salaries to teach their classes to women.[50]

The organizers hoped that women would eventually
become regular students alongside their male counterparts.
But as was the case with many universities flirting with coeducation in the nineteenth century, Harvard's experiment did
not fully welcome women into its community right away;
indeed, the university's own president, Charles William Eliot,
and the Harvard Corporation were "deeply opposed" to the
idea of allowing women into Harvard.[51]

By 1894, however, the Annex experiment had become successful enough to warrant an official charter from the commonwealth of Massachusetts. With a new moniker reflecting
the maiden name of Harvard's first female benefactor, Radcliffe College opened its doors for what would ultimately be
a 105-year stint. Before merging with Harvard and morphing
into the Radcliffe Institute for Advanced Study (now the Harvard Radcliffe Institute) in 1999, the school would become
one of the most storied women's colleges in America.[52]

By the time Millie arrived in 1952, Radcliffe certainly gave
female students more autonomy and authority than in its earliest days. But it was still very much a place where women
could feel like a different species compared with how their
male peers were treated.[53]

According to Millie, Radcliffe at the time didn't offer any
science classes; students interested in science degrees had to
take classes at Harvard. For her, this was a feature, not a bug:

the opportunity to learn with some of the best physicists in the world was one of the principal reasons she was interested in Radcliffe to begin with. Yet due to a difference in respective honor codes, all Radcliffe students had to take their exams separately from the men. This meant that they couldn't ask clarifying questions, and they weren't notified if errors on a test were discovered.[54] Instead, all Radcliffe women, regardless of field, "had to take their exams together in the same room because their presence in a coed examination setting was thought too distracting for the men," Millie stated in 1999.[55]

Up until this point, Millie had been fairly oblivious to gender biases in the sciences, or at least in opportunities to pursue science and engineering careers. She had, after all, attended a prestigious all-girls' high school and went to a college where the majority of both the student body and the faculty—including her mentor, Rosalyn Yalow—were women. At Hunter, she had found that male students—mostly veterans working on degrees via the Servicemen's Readjustment Act of 1944, a.k.a. the GI Bill—were often outshined in academics by their female counterparts, so she never picked up a sense that women were disadvantaged. She even noted that there had been a fairly large percentage of women in math and physics, at all levels, during her Fulbright fellowship in the UK.[56]

Millie's year at Harvard served as something of a rude awakening on this issue. "I felt a little odd," Millie later explained of her year in Cambridge, "because women were still very, very much in the minority and in some of the classes, I was the only one."[57]

In one particular course, she noted, her professor couldn't be bothered to come prepared for class, and he somehow came to call on Millie, the only woman present, almost every single session to provide a summary of what had happened the class before—like a secretary asked to read back the previous meeting's minutes, only Millie was a student and no other student was regularly put on the spot in this way. This recurring task placed an enormous amount of pressure on her, as the material was far from a cakewalk, and she didn't always understand everything about the prior session.[58] "He meant well, I guess, but that only aggravated my feeling of self-consciousness," Millie told *Cosmopolitan* magazine in 1976.[59]

One of her contemporaries, fellow first-generation New Yorker and eventual US Supreme Court Justice Ruth Bader Ginsburg, who attended Harvard Law School just a few years after Millie's master's program, had a similar experience as one of only nine women in a class of some five hundred Harvard men. "You felt you were constantly on display," Ginsburg recalled in a 2018 documentary film. "So if you were called on in class, you felt that if you didn't perform well, you were failing not just for yourself but for all women. It also had the uncomfortable feeling that you were being watched."[60]

As she soldiered on in her course work, Millie considered a doctoral program at the University of Chicago, which at that time was one of the strongest in the world for cutting-edge physics. It was home to, among others, world-renowned Italian American physicist Enrico Fermi, a giant in the field who had led the creation of the world's first nuclear reactor—and had become a Nobel laureate in 1938 for work on induced radioactivity and the discovery of transuranic elements, or elements with atomic numbers greater than that of uranium.[61]

Millie was intrigued by the possibility of joining a department with someone like Fermi at the helm. So after a successful application, her next move was sealed: following work that earned her a master's degree in 1953, her science career would begin in earnest in Chicago.[62]

4 MEETING OF THE MINDS

On the afternoon of December 2, 1942, on a squash court underneath the west grandstand of the University of Chicago's former Stagg Field, several dozen scientists looked on as a colossal mound of jet-black carbon became an indelible part of history. Following years of theoretical work and painstaking physical preparation, some forty thousand blocks of highly purified graphite, weighing approximately four hundred tons, sat in a meticulously designed lattice structure in a corner of the squash court floor. Inside the blocks, nineteen thousand pieces of uranium oxide and uranium metal lay in waiting. The goal of the scientists' experiment that day: to show that humans could create a controlled fission reaction, a continuous chain of splitting atoms that would give off free energy.[1]

At precisely 3:25 p.m., with instruction from the master of ceremonies—University of Chicago professor Enrico Fermi, the brilliant Italian-born physicist of the US Manhattan Project who, along with his Jewish wife, Laura, had fled the Nazi regime after picking up his Nobel in 1938—the last of a series of cadmium control rods was carefully removed from the pile, and the world's first controlled, self-sustaining nuclear reaction successfully commenced. Like the harnessing of fire hundreds of thousands of years ago, the experiment marked the dawn of an entirely new era, one in which humans could

suddenly control an extremely potent natural force—for both
good and destructive purposes.[2]

Eleven years later, with the nuclear age firmly underway
and the Cold War heating up, Millie found herself, at twenty-
two years old, one of the new students within the University
of Chicago's world-renowned physics department. A number
of researchers who had been influential during the Manhattan
Project had by then left for other opportunities. Yet with Fermi
and other luminaries—including Nobel laureates Harold Urey
and Maria Goeppert Mayer (with whom Millie lived for about
a year as a boarder), as well as physicist Leona Woods, the only
woman present during the 1942 fission demonstration—still
working and teaching, Chicago continued to host one of the
strongest physics faculties in the world.[3]

The school's physics program was fairly small in those days:
Millie had earned a spot as one of just about a dozen new
students that year. She was also, it turns out, the only female
student in the department. Yet Fulbright and master's degree
notwithstanding, Millie felt herself not quite prepared as she
began her PhD, due mainly to her subpar high school training
and uneven opportunities to catch up at the Cavendish and
at Harvard. So at the start of her doctoral studies, she revis-
ited her pre–Hunter High playbook: she found through her
department a cache of old examinations, and she worked the
problems therein forward and back until she felt up to speed.[4]

Despite this added practice, the course work for first-year
PhD candidates was brutal—so much so that around three-
quarters of all entering physics students eventually dropped
out of the program. But Millie received an unexpected boost
from someone who would go on to become another of her
great mentors.[5]

At first, her acquaintance with the unflappable Enrico
Fermi, who had made crucial strides not only in the development of the atomic bomb but in particle physics after the war,

was a formal one. In taking his class on quantum mechanics, Millie got to know the teaching style of the great master as patient, inspiring, and mind opening.[6] With a slow, deliberate, accented voice that Millie described as "halting,"[7] Fermi expertly distilled complicated topics, so that anyone in attendance could comprehend them. He delighted in stripping concepts to their essence, and unlike more impatient professors who were absorbed in their own work, Fermi cherished the opportunity to review whatever he knew about a physical concept by explaining it to someone else. For this he clearly had a talent. In the way he presented the finer details of quantum mechanics, Millie explained, "any youngster could think, when they heard the lecture, that they understood every word."[8]

One key to the eminent physicist's clarity was the ban he placed on taking notes. Fermi demanded full attention, so he would prepare and dole out handwritten notes before his lectures, lest students be tempted to take out their pens or slide rules. "What was so impressive and amazing about it is that the lectures were very exciting, whatever the subject was," Millie said in a 2001 interview.[9]

And then there was the homework, which was always tricky but delightfully enlightening once students figured it out. At the end of every class, Fermi floated a seemingly simple problem to be solved as an exercise prior to the following lecture. These included questions like: Why is the sky blue? Why do the sun and stars emit spectra of light? And, famously, how many piano tuners are there in Chicago?[10] "You thought it was simple until you got home," Millie said in 2012, upon receiving the Enrico Fermi Award, a lifetime achievement award given by the US Department of Energy.[11] These types of questions were later known, collectively, as "Fermi problems" and are taught today in schools around the world, from kindergarten all the way to graduate-level courses, as examples of how to estimate and triangulate in search of an answer even without knowing all of the relevant—and seemingly necessary—parameters. Back when Millie was learning about such problems, all she knew is they were due by the next class, no more than a day or two away, and they took significant effort to compute. "I think we learned a great deal from him in the formulation of problems of physics, how to think about physics, how to solve problems, and how to generate your own problems," Millie said.[12]

Indeed, throughout her career, Millie credited Fermi, whose genius allowed him to excel in both theory and experimentation, with teaching her how to "think as a physicist."[13] A key concept behind the Fermi system, Millie often stated, was the idea of single-authorship research. Graduate students were expected to conceive of, carry out, and publish their thesis work more or less on their own, without the guiding hand of a more senior faculty member. This required them to work with others to develop a broad understanding of physics that they could then apply to a research topic they'd generate themselves.[14]

Fermi's connection with students didn't end in the classroom. He was well known for frequent interactions with young people and for being the rare senior faculty member who regularly integrated students into his personal life. "It was not beneath him to associate freely with students and to treat them as equals," said Jay Orear, a career physicist and graduate student of Fermi, in a book of remembrances about his adviser. "In fact, I think he enjoyed young physics students more than some of his older colleagues."[15]

For Millie, this integration began, quite literally, on her way to school. She and Fermi lived in the same general vicinity, and both were early risers who walked down Ellis Avenue on their way to the lab each morning.[16] "I had him for class first thing in the morning. And on my way, walking to school, I would see him. And he would cross the street and walk with me," Millie recalled in a 2007 oral history interview. "That's just being very friendly, and that made a long-term impression on me."[17]

According to Millie, whenever they met, Fermi would always select the subject of discussion and would never fail to energize and inspire her. "I was a very shy youngster and wouldn't think of suggesting the topic to Enrico Fermi," she said in 2013. "He would always ask questions about 'What if this and this and this were true? What if we could make this—would it be interesting, and what could we learn?'"[18]

Fermi and his wife, Laura, were well known for hosting monthly dinners at their house, with dancing afterward, and they always invited his students. "Fermi especially liked young people," noted Harold Agnew, a longtime physicist and one of Fermi's graduate students, in a remembrance published after Fermi's death. "The top floor of his Chicago house had a large room in which he would invite students to come and square dance."[19]

"I remember those dinners," Millie said in 2012. "Laura Fermi was a very, very good Italian cook." But more than the cooking, she said, "it was the ambiance and the friendliness in that household that really made us enjoy physics—it was something more."[20] That "something more" would inspire Millie later in her career to provide her own students with a familial atmosphere at the lab, at group luncheons, and at events at the Dresselhaus's home, where lines between student and professor were blurred a bit and kindred spirits enjoyed one another's company.[21]

Millie's acquaintance with Fermi would last only a single year. He learned in that final year of his life that he'd developed an incurable stomach cancer, possibly a result of exposure to radiation from his earlier work. He died on November 28, 1954. But he left an enduring impression on Millie that she carried for the rest of her days, especially in relation to public service and to training her own students.[22]

"Fermi had the most profound influence on physics teaching in the United States, and our graduate programs . . . are much fashioned from his way of teaching," Millie said in 2001.[23] She later added, "From him, I learned that we don't have to be leaders in every field, but we can use our understanding to see connections that others might miss."[24] The broad physical and scientific knowledge that Millie developed as a result of Fermi's system for teaching graduate students would in fact come to help her in numerous ways throughout her career, including several occasions when she had to make significant course corrections, with very little background in the areas into which she pivoted, and as a leader of national programs with diverse constituents.[25]

Overall, however, perhaps the grandest lesson that Millie gained from her mentor was an understanding of what it takes to be a great teacher and advocate. "The most important thing that young people need is the confidence that they can succeed," she explained in 2012. "That's what I work on. When I have students, I make sure they are able to formulate and solve their own problems. I will help them, if they come in and talk with me. And I make sure they receive training for their next job. I always felt Fermi and Rosalyn [Yalow] were interested in my career, and I try to show the same concern for my students."[26]

MAKING CONNECTIONS

In many ways, Enrico Fermi—physics doyen and esteemed mentor—was, toward the end of his life, the brightest light within the University of Chicago Department of Physics. This is, of course, saying a lot, considering other scholars there had received impressive awards and accolades for their groundbreaking research. But despite his outsized personality and influence in the physics world, Fermi was not, in fact, the group's immediate shepherd. Rather, from 1950 to 1956, the department chair was Andrew Werner Lawson, a physicist best known for his studies of materials under high pressures.[27]

When it came time for her to select a research focus, Millie decided to place her attention not on the subatomic physics that Fermi and others had made their names on, but instead on solid-state physics, a subfield that focuses on larger-scale properties of solid materials. According to Millie, Andrew Lawson was, at the time, the only University of Chicago adviser in this area, and he was eventually assigned to oversee her as a doctoral adviser. Their student-adviser pairing was a mismatch from the start. It led to some unfortunate consequences for Millie, but it also forced her to become fiercely independent in her work, a circumstance that would come to benefit her in the long run.[28]

Lawson may have been an excellent researcher, but as of the mid-1950s, he held a deep-seated bias: he thought women should have no role in science, and he repeatedly told Millie as much. "He didn't believe I should be doing what [I was] doing," she said in a 2002 interview. "He was very unhappy every time I got a fellowship or any kind of recognition. He said it was a waste of resources."[29] The insinuation, of course, was that those resources could have been going to a man who would make better use of them as part of his up-and-coming career as a scientist—which he believed Millie couldn't possibly be serious about pursuing. "He never talked with me because he didn't think women should be in science," Millie said of her adviser in a 2012 interview with the Kavli Institute. "When I sought him out, he essentially told me to get lost."[30]

Lawson's attitude was not uncommon at the time and in fact persisted among male academics for decades. In a 1976 *Cosmopolitan* article on women in engineering, an unnamed department head at "a prestigious, Midwestern institute" stated, "Maybe I'm old fashioned . . . but I think a lot of the technical education we're giving women today is going to be wasted. They'll get married, have children, and their period of productivity won't last more than a few years."[31]

Unfortunately, the unwelcoming atmosphere didn't end with Millie's adviser. "I got a lot of negative input from people who believed that a woman's place is in the home," she once stated. "I heard so much of this that I believed it."[32] "Another thing I remember as a student was all the nude pinup girls on the walls," she recalled in a 1999 interview. "These were things you had to put up with in a lab to get your experiment working. It was not an environment in which women felt comfortable. But if you couldn't handle that, you knew you couldn't have a career in science."[33]

As was typical with Millie, such roadblocks only caused her to redouble her efforts—to find a way to follow her interest, no matter what it took. With Fermi out of the picture and Lawson focused elsewhere, Millie began taking lessons, both literal and figurative, from students and other professors in the department.[34]

One of those individuals was Clyde Hutchinson, who helped to develop the use of magnets and microwaves to study matter. Hutchinson was highly regarded for his work in magnetic resonance spectroscopy, a precursor of magnetic resonance imaging (MRI), the now ubiquitous process used in medicine and other applications in which the atomic nuclei of cellular material react to high-frequency radio waves in a magnetic field, leading to high-resolution images of tissues and organs.[35] Hutchinson's work shared similar themes with the thesis topic Millie eventually chose to pursue—microwave properties of superconducting materials in a magnetic field—and his experimental work helped her establish both opportunities and methods for testing out her hypotheses.[36]

Another individual who had a critical influence during this time quickly became Millie's cheerleader in chief for the duration of her doctoral program at Chicago and would continue to do so for the rest of her life. In the mid-1950s, Gene Dresselhaus was a rising star in theoretical physics who had come to the University of Chicago from the University of California at Berkeley. Born in 1929 in the Panama Canal Zone—a former US territory encompassing the Panama Canal and several miles on either side—Gene had studied physics at Berkeley, first as an undergraduate, and then in a PhD program under Charles Kittel, a noted professor in condensed-matter physics.[37]

As he was wrapping up his doctorate in 1955, Gene published a flurry of research papers. Two were coauthored by Kittel and fellow Berkeley physicist Arthur Kip; one of these investigated semiconductors silicon and germanium, while the other looked at the interactions of microwaves and magnetic fields, similar to those in MRI. A third paper, on which Gene was the sole author, represented the culmination of his doctoral thesis and vaulted him into science wunderkind territory. It helped form the early theory behind a property of electrons called spin that would eventually lead to technological advances now known as spintronics. Within this work was an important discovery that spin, in certain circumstances, can affect the range of energies that an electron may have within a solid material—a phenomenon now known as the Dresselhaus effect in his honor.[38] "Gene was a brilliant theoretical solid-state physicist," says his son Paul, himself a physicist. "His PhD work was a major advance in the way solid-state physics was done."[39]

After finishing his PhD, Gene accepted a postdoctoral research position within the Institute for the Study of Metals at the University of Chicago. There, he continued his condensed matter research, but he also began teaching—which was how he made his acquaintance with the future Queen of Carbon.[40]

Here I pause momentarily to note that prior to meeting the love of her life, Millie had in fact been married briefly to a fellow physicist, Frederick Reif, a Holocaust refugee who had immigrated to New York, studied at Columbia and Harvard, and began his career with a faculty position at the University of Chicago. Their relationship was not something she discussed publicly, but she did open up about this period of her life in a 1976 interview.[41] "I learned a lot about myself during those years," she said. "I learned about what was important in my life and what I wanted to do. . . . And I set priorities. I decided that a family was important to me; these things had a lot of impact in the next 10 years of my life."[42]

Parting ways with Reif, Millie would eventually meet the man who would become her greatest champion and lifelong companion. According to family legend, Gene freely admitted that teaching wasn't his forte, and he could be somewhat nervous in front of students. But his weakness proved to be a strength in the long run. One day after giving a lecture, the story goes, Millie came up to Gene and told him how bad he was at lecturing. This, according to Gene, was when he knew he was going to marry the young physicist before him.[43]

According to Millie's family, the two went on many dates—bike rides, walks, driving lessons for Millie, and the like. Gene was clearly smitten, and the feeling on Millie's end was mutual. In one letter to his mother, Gene didn't bother to hide his joy regarding his new companion: "Mother, when Carl [Gene's brother] finishes painting your place, why not send him to Chicago to paint an apartment for me. Millie has offered to help me, and she is better looking than Carl so on second thought you can keep Carl."[44]

EXPLORING SUPERCONDUCTIVITY

Millie had already begun the early stages of her doctoral research by the time Gene Dresselhaus arrived at Chicago, and he provided a flood of encouragement in the absence of a proper adviser. It took some time to set up her experiments, as Millie found herself working in a burgeoning discipline. At the time, American physicists John Bardeen, Leon Cooper, and Robert Schrieffer had yet to publish their Nobel Prize–winning theory of superconductivity, the tendency of chemical elements to lose all electrical resistance when cooled to a low-enough temperature. Fortunately for Millie, the University of Chicago had for the taking loads of surplus materials from the Manhattan Project. Her laboratory was located under Stagg Field—in the same vicinity where Fermi and his collaborators had started the first controlled nuclear reaction just over twelve years earlier.[45] "It was quite radioactive when I was there," Millie noted, only half-jokingly, at an awards ceremony in 2012.[46]

In developing the materials she would need to investigate superconductivity, Millie became quite the engineer. She produced, among other things, liquid helium and wires of superconducting material, and she fashioned her experimental devices from microwave equipment leftover from the Manhattan Project.[47]

During the 1955–56 academic year, Millie also benefited greatly from a visitor to the University of Chicago physics department: Cambridge University physicist Brian Pippard, whom Millie had known when she was a Fulbright fellow. Pippard's main objective during his year in Chicago was to study the behavior of electrons in copper, but he was also highly interested in superconductivity, and his work inspired Millie in a number of ways.[48]

In reading Pippard's earlier work on superconductors and their microwave properties, Millie began to formulate her thesis question: What would happen to the microwave properties of superconductors when they were placed in a magnetic field? Millie knew that when a superconducting material is put into a high enough magnetic field, its superconductivity disappears and the material transitions back to normal behavior. She wanted to observe what specifically happens in the midst of such a transition, as superconductivity goes away and the material becomes a regular conductor. Using her own handmade equipment, she experimented with a number of materials, tin in particular, as well as various transition temperatures and magnetic field orientations to see what she could learn.[49]

Her results produced a small wave in the physics community when she discovered an anomaly in the transition away from superconductivity. Under a magnetic field, it turned out, superconductivity in her test materials would actually increase a bit before dropping off completely.[50] "I saw it under many conditions, and it always seemed to be there," she recalled in 2001.[51]

Millie's work coincided with research that ultimately led to a Nobel Prize for its authors. In December 1957, as Millie was finishing up her thesis work, John Bardeen, Leon Cooper, and Robert Schrieffer published their landmark theory of superconductivity, known henceforth in the physics community as "BCS theory," the letters representing the researchers' surnames. While the theory does not apply in all circumstances—it does not, for example, explain materials that become superconducting at relatively high temperatures—it does explain how electrons can move inside a solid without resistance by grouping into pairs known as Cooper pairs. BCS theory has also been successful in providing groundwork for the development of applications such as the specialized magnets used in particle accelerators.[52]

When Millie heard about the BCS paper, she noted right away that there was no mention of the anomaly she had observed. Several months later, she reported her findings at a meeting of the American Physical Society. This immediately attracted some interest, including from John Bardeen and

Robert Schrieffer.[53] "Bardeen was actually very interested in my results, because [they] couldn't be explained by his theory," Millie noted in 2001.[54] Bardeen, who had only just won a Nobel Prize for his coinvention of the transistor, was in fact so intrigued with the inconsistency Millie had found that he invited her to give a colloquium at the University of Illinois at Urbana-Champaign, where he was a professor.[55]

The attention to Millie's thesis proved a boon to her career. Among other things, her experiments were replicated many times over—with slight variations. And while the cause of the anomaly she'd witnessed was ultimately found to have been related to electromagnetism rather than to superconductivity, the problem remained an open question for many years, with both Bardeen and Brian Pippard assigning students and others to look closely into the matter. "There was some good work that was done, and the electrodynamics of BCS was worked out" thanks to her findings, Millie later explained.[56]

Fourteen years later, in 1972, Bardeen, Cooper, and Schrieffer shared a Nobel Prize in Physics for the BCS theory, which by then had been further developed and enhanced by a number of colleagues. The honor marked the first time Millie would play a supporting role, albeit a rather small one, in the development of Nobel-worthy work.[57]

Yet despite her strong doctoral work, Millie still wasn't sure she could make a career of physics. "Women didn't have a lot of opportunities for careers in science" in the 1950s, she later noted. "Even though I had a good thesis, it was questionable whether I had much of an opportunity to go from there."[58] But Gene Dresselhaus, who had by that time left Chicago to become an assistant professor at Cornell University in Ithaca, New York, encouraged her, stressing that she could make it work. (He also, around the same time, persuaded her to marry him.)[59] "My husband Gene . . . was very important in convincing me that I could do physics and be successful at it," she said.[60]

Her official thesis supervisor, Andrew Lawson, who had been adamant that women didn't belong in science, was, according to Millie, completely unaware of her work until two weeks before she handed in her thesis. Nevertheless, her defense was a success, and she remained on track to complete her doctorate. (Millie would later note that some twenty-five years after her graduation, Lawson sincerely apologized for his earlier attitude and even invited Millie to present a distinguished lecture. "It was a very gracious thing to do," she said.)[61]

On May 25, 1958, Millie and Gene married in a simple ceremony. They were thrilled to start new lives as a couple and hoped to continue in physics together as well. Millie had secured a prestigious postdoctoral fellowship sponsored by the National Science Foundation, and she had some flexibility in where she could pursue it. With Gene at Cornell, she accepted an offer there for a two-year postdoc, but it remained to be seen what opportunities, if any, would surface after the fellowship was up.[62]

That summer, Millie was off for New York. The time Millie and Gene spent at Cornell was to be but a brief chapter in their history, marked with both immense joy and significant frustration—a defining period on multiple fronts.[63]

RUNNING IN PLACE

In later decades, Millie recalled achieving fairly little during her postdoc years. She published research from her thesis in *Physical Review Letters* in August 1958, right as she was moving to Cornell, and in *Physical Review* in March 1959, just before she returned to Chicago to pick up her diploma. Millie and Gene did accomplish some important follow-up work to her superconductivity research with magnets and published a handful of papers together during their time together at Cornell.[64] But overall, Millie considered the years from

1958 to 1960 "pretty disappointing" in terms of new work produced.[65]

At Cornell, Gene had been reunited with physicist Albert Overhauser, whom he had known from his graduate school days at UC Berkeley. When Millie began her postdoc in the summer of 1958, she too joined up with Overhauser and his colleagues in what was by all accounts a happy academic alliance. But the collaboration was short-lived: Overhauser left for Ford Research Laboratories just months after Millie's arrival.[66]

The Dresselhauses would later note that Cornell "lost its attraction" following Overhauser's departure.[67] Among other things, in losing a supportive colleague, Millie would once again be affected by attitudes about women in the workplace that might as well have come from a physicist's version of *Mad Men*—mocking at best and downright dismissive at worst. One faculty member told Millie that neither she nor any other woman would ever teach his engineering students—never mind her standout academic record and strong research experience, or the support she'd gained from the National Science Foundation to be there.[68]

"We were all supposed to publish papers, but for most people, they had some person that they could talk to," Millie said in a 2009 interview in *ACS Nano*. "The men who were doing condensed-matter physics at the time as I did, they could talk to this advisor who was available, but I didn't want to hear that I shouldn't be in physics [due to being a woman], so I didn't go talk to him. I did it by myself."[69]

Such misogynistic attitudes would be put to the test when a professor who had planned to teach a course on electromagnetic theory abruptly left Cornell, leaving senior faculty members scrambling to fill the teaching slot. Millie was already getting paid through her fellowship and was highly qualified to teach the course through scholarly familiarity with the subject matter and experience as a teaching assistant during her graduate student days at Chicago (not to mention years of tutoring from her high school and college days). She volunteered to take on the class.[70]

"There was a big uproar," Millie later explained. "The faculty met every day for a whole week to decide not whether I was qualified to teach the course, but whether the young men would pay attention to me. . . . It was difficult for these senior faculty to comprehend and deal with having a young woman teach young men."[71]

Millie went on to teach the course and, unsurprisingly, had a highly successful time of it. Years and decades later, students who had taken the class would come up to thank her for her unique insights and teaching style, which had made a big impression.[72]

During her postdoc years, Millie *was* productive in a completely different realm: in September 1959, she and Gene delighted in welcoming their first child and only daughter, Marianne. In what was most likely a combination of 1950s' expectations of a young working scientist—a societal model that generally assumed scientists were exclusively male—as well as her own personal drive, she doted on her new daughter but took almost no time off after she was born. "I don't remember Cornell, myself, but my mom always told me about how she took me into the lab with her—when I was only weeks old—and would do her work with me by her side," Marianne wrote in a 2018 remembrance of her mother. "When she went off to a lecture, the secretaries looked after me."[73]

For other women in Millie's position, Cornell might have been the last stop on a brief scientific flight aborted shortly after takeoff because of her husband's career taking priority. Fortunately for Millie—and for the rest of the world—she was not like other women, both in her own temperament and in having found a life partner who was at once a highly

sought-after theoretician in her same field and a fervent supporter of what he clearly saw as his wife's life calling.[74]

However, practical realities did set in as Millie's postdoctoral funding began to run out. The couple would soon have
to make some critical decisions due to a burdensome societal
challenge. In the late 1950s and early 1960s, antinepotism
rules prohibiting the hiring of two or more members of the
same family were firmly in place at many academic and other
research institutions, including Cornell.[75]

Millie and Gene considered that Millie might need to
accept work without compensation while Gene's salary as a
tenure-track professor covered the family's bills. However, it
soon became clear that there was simply no place for Millie as
a scientist at Cornell. "I offered that I would do physics and
not be paid . . . but they didn't even want me to do volunteer
physics," Millie recalled in 2009.[76] With no obvious opportunities for finding a relevant job elsewhere in Ithaca or the
surrounding area, the couple faced the following reality: "I
could either have settled down and become a housewife and
forgotten about my career, or we would have to move," Millie
recounted in 1976.[77]

Gene felt strongly that he couldn't possibly let his partner's talent go to waste while he enjoyed a traditional career
as a physicist simply because of their respective genders. He
assured Millie that they would find some way to both work
in science, even if it meant giving up his highly coveted faculty position. "To his great credit, he wanted to go someplace
where both would be happy," Marianne Dresselhaus Cooper
says of her father.[78] Elizabeth Dresselhaus, one of Millie and
Gene's five grandchildren, affirms, "I think that Gene believed
in Millie's potential, in a way that few people would believe in
a woman aspiring to be a scientist at that time."[79]

"Gene recognized [Millie's] unique genius," adds Laura
Doughty, the Dresselhauses' longtime administrative assistant

at MIT. "He was no slouch in the physics world. But because he was so smart, he recognized that she really had the goods. And so, he saw his role as doing as much as possible to clear her way and enable her to do the work that only she could do. . . . Gene was always fiercely protective of Millie, and he didn't want her to be anywhere where people treated her badly."[80]

It didn't take long for Gene and Millie to find employers eager to hire them both. In casual conversations with friends and colleagues at a meeting of the American Physical Society, they floated the idea that they were looking for jobs and were immediately courted with offers. The Dresselhauses had decided it would probably be best to focus on research rather than academic jobs because of the immense amount of additional work they'd taken on caring for a new baby—and, looking ahead as they planned to expand their family further, they knew they'd have precious little time to develop new curricula, teach classes, and grade papers each week.[81]

While a number of labs conveyed interest, two outstanding research institutions extended offers that met their main requirement: the ability to work in the same place without restrictions related to nepotism rules. Gene and Millie had a difficult time deciding which to pursue, and like anyone stuck on the horns of a dilemma, they weighed the pros and cons: "We made up a score sheet, and we put down a whole list of conditions . . . and then we rated them," Millie said.[82]

One of the two finalists was IBM, the world-renowned technology firm that by 1960 had decided to go all-in on basic science and engineering research in order to support its development of next-generation computing machines. The company's first basic science research lab had been established in 1945 at Columbia University on the Upper West Side of Manhattan; later, in response to the invention of the transistor by John Bardeen, William Shockley, and Walter Brattain at Bell Labs, IBM's Poughkeepsie, New York, laboratory began focusing on solid-state physics and engineering research. From 1950 to 1954, the number of IBM employees in research and development jumped five-fold, from six hundred to three thousand.[83]

By 1956, IBM executives had decided that the company would consolidate its research efforts and separate them from development so as to draw the best in the field. "Among the anticipated benefits of a separate research organization, with greater emphasis on basic research, was the ability to attract outstanding college graduates with advanced degrees in science and engineering," wrote Emerson Pugh in his 1995 book, *Building IBM: Shaping an Industry and Its Technology*, adding, "Typically these young people were imbued by their professors with the belief that product development was an inferior activity to basic research."[84]

Just as Millie and Gene were about to leave Cornell, IBM was putting the finishing touches on its Thomas J. Watson Research Center, a gleaming new research headquarters in Yorktown Heights, New York, a wooded bedroom community about an hour north of New York City. The directors at IBM extended opportunities to both Millie and Gene to join the company in its quest to unlock new insights into the nature of materials and to use that knowledge in building its machines of the future. Critically, they were happy to have the Dresselhauses not only on the payroll at the same time but working directly together if that was what they desired.[85] "They really wanted to have us," Millie said.[86]

IBM's best competing offer came from a research facility on what is now the Hanscom US Air Force Base, situated at the confluence of four Boston suburbs: Lincoln, Lexington, Bedford, and Concord. Established by MIT in 1951, Lincoln Laboratory is a federally funded research and development facility that was founded at the start of the Cold War. It came into being on the heels of a 1949 report, shocking most in the

United States, concluding that the Soviet Union had secretly exploded its first atomic bomb on August 29 of that year. The United States at the time was effectively naked against a nuclear attack, so the Department of Defense went into a rapid phase of strategizing.[87]

After an ambitious investigation by George Valley, an MIT physics professor, a full air defense research laboratory, funded by the US government but administered by MIT, was proposed. MIT researchers had collectively become a world authority in radar during World War II and had already developed some of the computing technologies that would be needed for a nationwide air defense system—including Whirlwind, the computer that Millie had passed up the opportunity to work on when she accepted her Fulbright fellowship. Following numerous necessity and feasibility studies, the laboratory was established as Project Lincoln. It began with a staff of several hundred, but within two years, the number had skyrocketed to about 1,800.[88]

By 1957, the lab's initial charge—providing the research to create a functional air defense system—had officially been achieved with the development of the Semi-Automatic Ground Environment (SAGE) air defense system. Testing and implementation were considered by many to be outside the scope of the lab's work, and at that point MIT administrators began rethinking the lab's role in military applications. Meanwhile, in October 1957, the Soviets launched Sputnik 1, the first artificial Earth satellite. By the time Millie and Gene began looking for work a couple of years later, Lincoln Laboratory had begun to branch into new research areas affecting national security—and some of the scientists and engineers working in these areas would inevitably need to learn the latest in solid-state physics.[89]

Both Millie and Gene were offered staff positions in the Lincoln Laboratory Solid State Division headed by physicist

Benjamin Lax. Beyond the benefit of allowing wife and husband to work closely together, the offers were generous in a
unique way: after certain requirements to help anyone at Lincoln who might need support in understanding condensed-
matter physics for their defense work, the Dresselhauses could
use the rest of their days pursuing high-caliber basic science
in (mostly) whatever areas they liked.[90] "What could be
better than that?" Millie exclaimed in a 2007 MIT oral history. "It was kind of an open-ended job. Those kinds of jobs
don't exist today."[91] After considerable deliberation, Millie
and Gene decided that Lincoln Lab was their choice. All that
was left to do was find a place to settle down, a nest for their
growing family.

In September 1960, the Dresselhauses moved into the two-
story, four-bedroom house that they would call home for the
next fifty-five-plus years. It was conveniently situated about
halfway between Lincoln Laboratory and the main MIT campus in Cambridge, where Millie was already conducting experiments. She told the *Arlington Public News* in 2015, "We walked
in there, it looked just absolutely fine, price was right, people
were very welcoming, we moved right in, and that was it!"[92]

Some fifty-eight years later, on a drizzly April afternoon,
Marianne Dresselhaus Cooper welcomes me into her parents'
home in suburban Arlington, Massachusetts. The eldest of
four siblings, she bears a clear resemblance to both Millie and
Gene: broad smile, wise blue eyes, and light, wavy hair pulled
back into a ponytail. As I look around to take in the house
that the Dresselhauses lived in for over half a century, it feels
cozy and intimate. While this part of town is quite desirable—
with easy access to Cambridge and Boston and an idyllic 36-
acre park abutting their backyard—the house seems modest
for what was, for many years, a family of six (plus a longtime
babysitter and caretaker who was there during the day).[93]
Mementos from Millie and Gene's lives in science abound:

framed photos of colleagues and friends from bygone eras, mugs emblazoned with logos of scientific conferences, and so forth. The central place of music in Dresselhaus family life is also on display: music stands and sheet music blanket the living room, evidence of the parties, Thanksgiving dinners, and countless musical guests they hosted, Millie on violin or viola, Gene as conductor.

One can easily imagine a far more lavish homestead that any scientific power couple might have set up for themselves. But knowing Millie's difficult beginnings and humble outlook, this charming home makes perfect sense. When she and Gene chose it in 1960, they were just beginning the main sequence of their research careers. Throughout the decades, it would serve as an incubator for many aspects of the Dresselhauses' lives: a place to relax and to raise family, of course, but also a place where ideas would evolve and flourish, where students and associates would trade insights and bad jokes, and where echoes of Mozart and Schubert and Brahms can still be heard.

5 A SCIENTIST BLOSSOMS

Even as a young student, with his wide smile and signature protruding ears, physicist Benjamin Lax was both thoughtful dreamer and gifted tinkerer—the kind of scientist who was fascinated at once by theories of nature and by the thrill of executing empirical tests of his hypotheses to find out whether they held any water. His PhD work at MIT, on the effect of a magnetic field on plasma—gas with excess charged particles that joins solid, liquid, and gas as a fourth state of matter—was partly theoretical and partly experimental, and it presaged the research he would both undertake and oversee throughout his prolific career.[1]

Lax joined MIT Lincoln Laboratory at its inception and quickly developed a reputation for taking on new projects and ensuring they were successful. By 1958, around the time Millie Dresselhaus left Chicago for Cornell, he had vaulted his way into a prestigious management position. As head of the laboratory's Solid State Division, Lax and his colleagues focused on the development of so-called solid-state devices: electronics such as transistors, integrated circuits, and photovoltaic solar cells that were based primarily on semiconducting materials— substances such as silicon and gallium arsenide that can act as either a conductor of electricity or a nonconducting insulator, depending on how they were prepared and on their environment.[2]

In order to better understand the workings of semiconductors and their cousins, semimetals (elements that share some properties of metals and nonmetals), Lax and other researchers in the Solid State Division spent time some at the MIT campus in Cambridge in the basement of Building 4 on the eastern edge of the Institute's famed Killian Court. There, in a small laboratory founded and managed by MIT professor of physics Francis Bitter, the researchers had access to high-intensity electromagnets, which could be used to generate controlled magnetic fields. Employing the magnets in Bitter's lab, which could produce significantly stronger magnetic fields than the equipment at Lincoln, Lax and his colleagues could conduct experiments investigating the electromagnetic properties of various materials of interest.[3]

There were, however, limits to the magnetic field intensity they could achieve, and Ben Lax had visions of creating a much bigger and more sophisticated magnet laboratory, one with more than twice the magnetic field intensity available in Bitter's lab. Such a state-of-the-art facility would serve not only the MIT and Lincoln research communities but would also attract scientists and engineers from around the world.[4]

A highly motivated mover and shaker, Lax quickly succeeded in turning his forward-thinking vision into reality. He formed and led a team, which included Bitter, that secured the necessary funding from the US Air Force Office of Scientific Research to establish the National Magnet Laboratory—later the Francis Bitter National Magnet Laboratory, after Bitter's death in 1967—the first high-magnetic field facility in the world and, eventually, the best facility for anyone looking into properties of matter using intense magnetic fields.[5]

The National Magnet Lab was still in the planning stages when Millie and Gene Dresselhaus connected with Lax about possible work at Lincoln Laboratory. Lax had known both scientists through prior work; in particular, he had collaborated with Gene on influential experiments related to cyclotron resonance, a phenomenon in which charged particles in a magnetic field are exposed to external forces.[6] Unlike many other research administrators at the time, Lax was enthusiastic about employing women. Several years earlier, he'd recruited physicist Laura M. Roth to work in his group. After Millie joined, he would, according to Millie, tell visitors he was working with "the two best young women physicists in the country."[7]

When the Dresselhauses began their new jobs, they were told that the direction of their research was wide open, with one key exception. Lax, whom Millie later credited with giving her and Gene great research freedom, held a suspicion that superconductivity, Millie's specialty, was essentially finished as a science subject, with the biggest advances—in particular the 1957 BCS theory that would eventually earn its developers a Nobel Prize—already discovered.[8] "He was more interested in taking advantage of masers and lasers, which had just been invented, to study the properties of semiconductors," Millie recalled of her former boss in a 2012 interview.[9] Lasers (an acronym for light amplification by stimulated emission of radiation) and the related masers (microwave amplification . . .) were then just baby technologies, and researchers at Lincoln Laboratory and elsewhere were busy dreaming up new applications.[10]

Lax ultimately directed Millie, for her work outside of advising on staff projects, to focus on any area other than superconductors. Considering she was recruited to join Lincoln Laboratory because of her stellar research on superconductivity, it would have been natural to be frustrated at the change of direction. But in typical Millie fashion, she simply considered the pivot as a great opportunity to learn another area of physics—something that came naturally to her based on her go-with-it attitude as a child in search of opportunities in New York City and on the important lesson that Enrico Fermi had seared in her mind during graduate school: that being able to change directions and have a broad understanding of science rather than a myopic and more limiting one was by far the more enlightened—not to mention practical—path.[11] The mandate "turned out to be a big positive in my career, because I learned another field," Millie said in a 2007 oral history interview. "This worked out really well for me."[12]

In addition to lasers and masers, Lax was interested in a physics subfield known as magneto-optics, which, loosely defined, concerns the interaction of light and materials in a magnetic field. Intrigued, Millie took a closer look. She had previously worked with magnetic fields in her superconductor research. But the addition of an optics element was new and would require learning a completely different experimental technique, which would take a considerable amount of time and effort.[13] "The idea was to induce a magnetic field and use lasers to see how the electrons behaved in high magnetic fields," she later explained.[14]

Millie spent the next six months or so learning the particulars of magneto-optics. To begin, she worked strictly with semiconductors, those situation-dependent, electricity-conducting materials that many researchers were interested in for their applications to computing. In fact, Millie said, most of her colleagues in the Solid State Division at the time were focused on semiconductors and how they worked. Those specializing in magneto-optics hoped to better understand the ranges, or bands, of energies that an electron within a solid semiconducting material can occupy. This knowledge would provide researchers with a firmer grasp on the electronic properties of those materials, which could be used to develop next-generation electronics.[15]

Millie, however, never caught the semiconductor bug. She was intrigued at first, but her interest waned after she studied a couple of semiconducting materials. She also wasn't thrilled with the fact that so many of her colleagues were working on the same general problem.[16] Ever inquisitive, Millie wanted to research something unique. And so, after her magneto-optics boot camp, she decided to apply her newly acquired foundations toward other elements: semimetals, starting with bismuth (Bi), atomic number 83.[17]

An extremely dense and very slightly radioactive element that naturally forms iridescent stair-step crystals, bismuth often features a psychedelic multicolor oxide tarnish—making small chunks of it look like something you'd see boxed up in the Willie Wonka factory. The electronic properties of bismuth had intrigued researchers in the Lax group, and experiments to understand its cyclotron resonance were already underway when the Dresselhauses arrived. Millie and Gene joined the project, and when experimental work moved to the basement of MIT's Building 4 in order to take advantage of Francis Bitter's higher-power magnets, Millie got her first taste of the MIT campus.[18]

The project yielded promising initial results, but a conflict within the research team soon presented a challenge for Millie: a colleague with what she called "a difficult personality" seemed not to want to work with her—possibly because she was a woman, though she didn't know for sure.[19] Millie eventually decided to quit the project and find a new one. "Nobody made a big issue of it," she later explained. "I had too many things to do, and if he wasn't happy having me, I was happy doing something else."[20]

In search of a new research direction once again, Millie consulted with her indispensable science sounding board and partner, Gene, who suggested she move into carbon, a wonder element found in countless compounds. In his previous studies, Gene had worked on various carbon allotropes, including graphite. From this work, he had formed the opinion that carbon was hiding some rather fascinating physics. It didn't take long to convince Millie that carbon's electronic properties had the potential to be uniquely interesting—and that it would be worth spending the time to investigate them more fully.[21] "Here was a material that had properties like a semiconductor, but it wasn't a semiconductor at all," Millie noted in a 2015 interview.[22] In particular, she was curious about the electron energy bands that appeared to be unique to carbon and had yet to be fully explored.[23] "The little bit I learned made me wonder why no one was interested in it."[24]

One reason that Millie and Gene's colleagues had thus far avoided a material like carbon was that they thought its atomic properties were too labyrinthine to study properly—and unlikely to yield important results when navigated.[25] Carbon was noted for having four distinct electron energy bands—two more than semiconductors. When it came to understanding a material's electronic properties, additional energy bands added to its complexity.[26] "My bosses at Lincoln Lab were skeptical and told me that they did not expect me to make much progress with this project," Millie said in 2002. "I thought this was a good challenge."[27]

With support from Gene, Millie came to see quite a few upsides to studying carbon. From a science perspective, she realized the carbon system featured certain properties that would make it quite attractive as a research subject. One was its small electron effective mass—the mass an electron in a material appears to have in response to external forces: the smaller the electron effective mass, the easier it is to study the electrons in a material. Another was that the energy levels that electrons could inhabit within a carbon atom were spaced far apart, which would make them far more tangible for research purposes than if they were closer together.[28]

Millie was also happy to tackle carbon for personal reasons. She appreciated, for one thing, that competition was practically nonexistent. "I decided I wanted to have projects that were a little bit slower moving that I was unique [to], and I didn't want to have to compete against the whole world," she explained in a 2001 interview.[29]

Lighter competition, of course, meant that Millie wasn't expected to produce regular groundbreaking results. This was a huge side benefit during this period of her career because for much of the early 1960s, she was either pregnant, raising babies, or both. "In society at that time, it was believed that women should not have children beyond 35 years old," she stated in 2009. "I hurried up and had all the children I was going to have before 35."[30]

In the span of five years, from 1959 to 1964, Millie and Gene had four children in quick succession. Their second child and first son, Carl, arrived in January 1961. "I never knew when I would have time for serious work," Millie noted in 2002. "This is why I wanted to have some problem that was sufficiently hard, but didn't have too many people working on it; a topic that wasn't too attractive to a lot of people."[31]

Millie would benefit greatly from a dedicated nanny for the bulk of her child-rearing days—a woman named Dorothy Terzian, who worked for the Dresselhauses for some thirty years, tending to the children, preparing family meals, and otherwise running the household in Millie and Gene's absence.[32]

"Millie would always say that [hiring Terzian] was the reason she was able to do all that she did," son Paul Dresselhaus says. "She'd already raised her five children and was a wonderful influence on us. In those days, women were expected to stay home with the kids. Millie said that when she started, she was paying Mrs. Terzian more than she made, but it was an investment that was necessary for her career."[33]

Despite Terzian's enormous assist, Millie admitted that the earliest period with her young ones was the most intense in terms of negotiating work-home balance. As a result, taking on carbon, which (almost) no one else wanted a piece of, provided a bit of a buffer.[34] "If one day I had to be at home with a sick child," she explained to the *New York Times*, "it wouldn't be the end of the world."[35]

Even so, this was a time before parental leave became a common employment benefit, and Millie famously took a combined total of five days off work following the births of her three sons—one of whom, she later noted, was born on a snow day, while another joined the world on a long holiday weekend.[36] "I was sufficiently interested in my work that I couldn't put it down at all, so I even took my briefcase to the hospital with me when I had babies," she said in 1976.[37] Awkwardly, when she returned to work a day or two later, her colleagues sometimes didn't even realize she had just given birth: "When people would ask me, 'When are you going to have your baby?' I said, 'I already had it! Don't I look different?'"[38]

At the time, Millie and Laura Roth were the only two women employed as scientists at Lincoln Laboratory; all other staff scientists, about a thousand in total, were men. At MIT, where Millie carried out experiments, only 2 percent of students were women.[39] "We were," she told the *New York Times*, "pretty much invisible."[40]

To be sure, Millie knew that establishing a new field was "likely to be a lonely venture" for anyone:[41] Papers might go unread for years—or, worse, they could end up in the dustbin of scientific enterprise. Conferences don't yet exist to help those with interest in the new area connect, share findings, and brainstorm nascent ideas. The silver lining to this inevitable lack of niche collegiality was the knowledge that those contributing at the start of a new subfield had an outsized opportunity to influence future generations of scientists once the field took off.

In Millie's case, the following fifty years of her life were proof that she and Gene had made the right call. "The number of papers published on carbon when I started was essentially zero, and it's been going up, up, up my whole career," she told the *New York Times*.[42] What's more, she noted proudly in her 2007 MIT oral history, it was these carbon-based studies that "turned out to be the important work."[43]

But beyond her ultimate historical legacy, carbon in the 1960s became the spark she had been searching for—a flicker that ignited a long, slow burn of continuous progress in a field that, true to Gene's words, was full of promise for one as curious as Millie.

GETTING TO KNOW GRAPHITE

Were it not for high school chemistry—or books like this one—you might never realize that the brilliant rocks strung together in a diamond necklace and the dull blackish filling of an everyday pencil are elemental siblings, made of essentially the same substance—solid carbon—but drastically different in appearance and behavior (figure 5.1).

Diamond, from the Greek *adámas* (unbreakable), is composed of crystallized carbon in which each atom shares electrons with four other carbon atoms. The bonds keeping this crystal lattice together are notoriously difficult to break, which is why diamond is among the hardest-known materials on Earth and commonly used in industrial-strength cutting, drilling, and grinding machinery. (Diamonds are not, incidentally, inherently rare as a mineral; they are expensive mainly because their supply has been deliberately restricted, principally by a single company and because this same company has

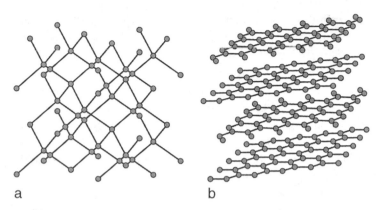

a   b

FIGURE 5.1

In diamond (left), each carbon atom is strongly bonded to four other atoms. In graphite (right), carbon atoms are strongly bonded to three other atoms in (virtually) two-dimensional planes. In-plane bonds are incredibly strong, while bonds between planes are weak.

successfully cultivated a marketing strategy that manufactured the perceived need for diamond's shimmer—as a symbol of affection and of wealth or social stature.)[44]

Meanwhile, pencil lead—not to be confused with elemental lead (Pb), a heavy metal—is made chiefly of graphite, another form of naturally occurring carbon. Unlike the tetrahedral lattice of diamond, soft graphite comprises carbon atoms arranged in hexagonal rings within two-dimensional sheets known as graphene. These sheets are weakly bound to each other through so-called van der Waals forces, which makes graphite both supremely tough and incredibly slippery. In-plane bonds holding carbon rings together within graphene sheets are exceptionally strong—one of the strongest found in nature—while between-layer bonds can be broken by any child whenever she scribbles with pencil on paper. I like to compare graphite's structure to the many layers of dough of croissants, puff pastry, and other flaky baked goods. Phyllo, for one, gets its name from the Greek for leaf and is the basis for baklava, a mainstay of Middle Eastern cuisine—and a delicious metaphor for understanding bulk graphite.[45]

In pencils, graphite is mixed with a bit of clay for improved manufacturing and greater variability of darkness in written or drawn marks. But pencils aren't the only place to see graphite in action. This form of carbon is also in batteries, nuclear reactor parts, machine lubricants, and carbon-enforced plastics, from tennis racquets and bicycles to panels of airplanes and sports cars.[46]

Graphite is an electrical conductor, so anyone can use a pencil to complete an electric circuit. A millimeter-thick sample of graphite—around the same diameter as the lead in an average pencil—contains about 3 million layers of graphene, and those layers have proven especially intriguing.[47] As described by the Nobel Committee for Physics in relation to the 2010 physics prize, awarded to Andre Geim and Konstantin Novoselov "for groundbreaking experiments regarding the two-dimensional material graphene"—which, as we will later see, had been studied theoretically by Millie and others decades earlier—graphite's layers are nearly transparent yet so dense that no gas atoms can pass through.[48]

Millie wanted to understand everything she could about carbon's fascinating properties and to tease out the differences between some of the element's various forms. She set about doing this around the same time that construction on the National Magnet Laboratory began at MIT in a space that had formerly served as a bakery. Ben Lax, the facility's founding director, had initially hoped the new magnet lab would be built at Lincoln Laboratory, but Cambridge was ultimately selected, in part due to the proximity to many research universities in the area.[49]

During this time, Millie was spending about half her days at Lincoln, completing her advisory tasks and connecting with colleagues, including Gene, about her research. The other half of her time was spent at MIT, in the basement of Building 4, tinkering with magnets.[50]

"In the 1960s we were busy trying to understand graphite as a bulk material, the very simplest form, when you stack the layers together as they are in nature," Millie recounted in 2013. "There was a theory for that, but there were very, very few experiments that shed any light on whether that theory was accurate. . . . We wrote many papers in many aspects measuring any conceivable property we could think of."[51]

Millie's first experiments were disappointing. She tried various graphite samples, which at the time were produced by industry for military and other government applications such as the space program. But they were not pure enough; defects in the crystalline structure were coloring the results of her experiments in a way that rendered them useless.[52]

Thankfully, the emergence of a synthetic graphite that was extremely pure and free of defects solved Millie's problem in relatively short order. This so-called highly oriented pyrolytic graphite, developed at Imperial College in London and produced by General Electric, enabled a whole new era of carbon research using magneto-optic techniques.[53] Once Millie got her hands on a sample, she knew she was in business. "The first time we tried it, it was beautiful," she said in 2001.[54]

Over the next couple of years, Millie collected reams of data relating to experiments with synthetic graphite. Her goal was to translate observed light spectra, which hid detailed information about the electronic structure of carbon, into a meaningful model. It wasn't easy for Millie and Gene to interpret what they were seeing in the data. At the time, this line of inquiry represented wholly uncharted waters—so much so that in performing her experiments, Millie and her colleagues would be among the first scientists to use lasers to probe the behavior of electrons under a magnetic field.[55]

The Dresselhauses turned to a former classmate of Millie's from the University of Chicago for clues to what Millie had been observing. Joel W. McClure, a physicist at the University of Oregon, had recently been working on a theory explaining the band structure of graphite.[56] "I really got a lot of seminal ideas from him," Millie said in 2001. "Gene understood immediately how to translate what [McClure] was doing to the experiments that I was doing."[57]

Thanks to this connection, the scientists were able to tease out some early details surrounding graphite's electronic structure. In the July 1964 issue of *IBM Journal of Research and Development*, Millie and McClure both published research in this area, Millie on the so-called Fermi surface of graphite—a concept that delineates what energies electrons may have in a solid material—and McClure on graphite's energy band structure.[58]

For Millie, the work represented an early step in a field that would be relatively slow to mature but that by now is booming. Today, understanding how graphene—a single, freestanding layer of graphite—can efficiently channel electrons has become one of the most pressing questions within materials science and engineering. But when Millie was studying graphite at Lincoln Laboratory, she and Gene were among just a handful of researchers to give any form of carbon the time of day. "There were three papers per year in the world, and I think they were almost all mine," Millie once noted in *MIT Technology Review*.[59]

Indeed, some of Millie's earliest work in this area was so innovative that it wasn't properly acknowledged until years later and was overlooked until others remade the same discoveries more than a decade later. "At the time we did it, people did not appreciate what we were doing," she noted in 2002. "I was young when I sent in [one] early paper and the work wasn't appreciated by the referee. I didn't fight to get it published; I said, OK, if it's not good enough, it won't be published. . . . One way or another many things did get published, but maybe not in the form that was initially intended."[60]

While it took quite a few years for interest to catch up, Millie and Gene's earliest adventures in carbon science set the stage for technologies that have already changed the world—such as carbon fiber composites that have transformed industries from aviation to athletics. They also laid the groundwork for new science and engineering that are just now revolutionizing technologies of the future, from flexible digital displays to quantum computers.

TAKING CHARGE

Let's take a brief science break to understand why determining the electronic structure of graphite presented such an interesting problem for Millie and Gene. At the heart of their early carbon research was the question of what it means for a substance to be an insulator, a semiconductor, a semimetal, or a metal—how it handles electric charges, in other words.

Within individual atoms, electrons orbit the nucleus in specific ranges known as orbitals. Each orbital has its own energy level, and the electrons of a given atom fill up various energy levels in a predictable, stepwise manner (figure 5.2).[61]

In a sample of solid material, trillions of atoms might be interacting together, sharing electrons. When individual atoms join within a solid, their atomic energy levels overlap to form a range of molecular energy levels, better known as energy bands. Within these bands, electrons can essentially move around at will.[62]

A material can have many bands, but two are particularly important in terms of understanding the material's electronic structure and conductivity (figure 5.3). These are the valence band and the conduction band, and the spacing of these

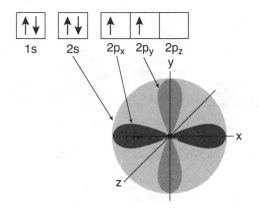

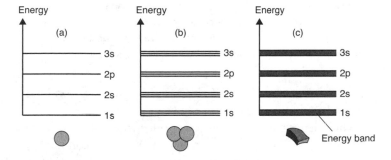

FIGURE 5.2

Top: Orbitals are regions of space surrounding an atomic nucleus where electrons are likely to be found. In this simplified view, we see the ground state of carbon, which contains six atoms. Each orbital can hold only two electrons. The 1s orbital, seen as a central dot, is enlarged so as to be visible. Bottom: Generally electrons fill atomic orbitals in a stepwise manner, according to the energy level of the orbitals. (a, b) Energy diagrams for a single atom and multiple atoms, respectively. (c) Energy bands form when electrons from multitudes of atoms overlap at discrete energy levels.

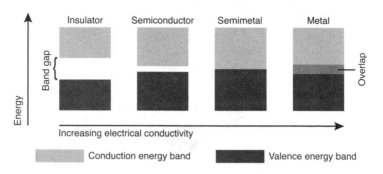

FIGURE 5.3

A material's electrical conductivity depends on the spacing of its energy bands. When bands touch, as in metals and semimetals, energy flows unimpeded. Energy cannot flow when bands are spaced too far apart, as is the case for insulators. But small gaps, such as those in semiconductors, can be overcome under certain conditions. The energy diagrams shown here are simplified.

bands is what determines if and how the material conducts electricity.[63]

The valence band is the outermost electron orbital in which electrons exist in a given atom. Meanwhile, the conduction band is the band into which electrons can move if they become excited enough. In some materials, there's an energy difference between the valence and conduction bands, known as a band gap; here, electrons are forbidden from hanging out. But under the right circumstances—such as the application of the right amount of heat or light—electrons can jump from the valence band into the conduction band. When this happens, an electric current is created, and the electrons are free to move to other atoms within the material.[64]

One might think of this band system as a scene at a popular nightclub. The valence band is the area where energetic clubgoers gradually fill up the sidewalk, waiting and hoping for a chance to get inside and party. The conduction band is the club, and the band gap is the bouncer restricting entry.

In metals (that is, conductors) the valence and conduction bands overlap such that it's no problem for an electron to flow freely between bands. In our nightclub analogy, the doors are wide open, there's no bouncer, and people can just head on in at will. Metallic cooking pots and flatware, metal wires and foils: all of these objects are good electrical conductors, made of materials in which electrons easily come and go.[65]

On the other side of the conductivity spectrum, insulators — such as plastic, wood, and rubber— don't conduct electricity well at all. The reason is the large energy gap between the valence and conduction bands, which acts to restrict the flow of electrons. In our nightclub, materials with the largest band gaps also feature . . . the largest bouncers! At least historically, large bouncers do have a reputation of being more effective at deterring especially eager clubgoers. Seeing our analogy through, the largest bouncers are also the best negotiators (from the club's perspective, anyway). They might let some electrons in, but they can also talk even the most well-dressed, well-coiffed, and highly bejeweled clubgoers into partying on the street, even in the face of some argument (in the form of heat, light, electric charge, or an electric field).[66]

Speaking of bling, most diamond crystals act as insulators. As we learned previously, diamond is a three-dimensional lattice of carbon atoms. When diamond's valence orbitals are completely filled, there's a large energy gap between the valence and conduction bands. Even at high temperatures (i.e. with the addition of heat), there's simply not enough energy to help electrons regularly jump up past the band gap and into the conduction band, so current can't flow.[67]

Semiconductors fall in between conductors and insulators in terms of their conductive properties. As with insulators, once electrons fill up the valence energy band of a semiconductor, they have to contend with a band gap before they can consider jumping into the conduction band and creating an electric current. But in semiconductors, including silicon and germanium, the size of the energy band gap—and the nightclub bouncer—is much smaller and therefore more easily surmountable. In addition, when heat, light, electric charge, or an electric field is applied (and who wouldn't talk a bouncer up a little when he or she is standing between you and the night of your life?), electrons are much more likely to move into the conduction band, creating electricity.[68]

We now come to the semimetals, a group of materials including graphite whose properties typically fall between those of metals and those of nonmetals. When Millie and Gene began investigating these substances in the 1960s, very little was known about them, as they didn't fall neatly in the metal, semiconductor, or insulator category. We'll dive deeper into some of the Dresselhauses' discoveries in later chapters, but for now we can say that semimetals like the graphite form of carbon have very slightly overlapping valence and conduction bands, meaning there is no band gap to overcome—and no bouncer to speak of at the nightclub. But unlike metals, semimetals have lower overall conductivity, so you might say that for the Semimetal Nightclub, there's a secret door that you have to know about in order to get in. That said, unlike semiconductors, semimetal conductivity is always nonzero, and electrons in graphite and graphene get to dance without too much fuss. But the differences in these and other semimetal materials took time to tease out.[69]

LEAVING LINCOLN

On top of their innovative work on graphite and semimetals including bismuth and antimony, the mid-1960s saw Millie and Gene extremely busy in their personal lives as their young family grew. In January 1963 their third child and second son, Paul, was born. The following year, when Millie was thirty-three, the couple's fourth child, Eliot, arrived in July.[70]

At that time, American families with four children were much more common than they are today—and by all accounts, Gene was an equal partner in raising the children. Nevertheless, things did begin to strain with unrealistic expectations of Millie as a mother and a working scientist, as on the day after she delivered Eliot, in 1964.[71] In a 2015 *IEEE Spectrum* profile, Millie's former Lincoln colleague Eugene Stanley recalled seeing her at work that day: "She was there around noon or 1 o'clock with the baby in tow. But because Lincoln Lab was a government lab, you either had to have clearance or have a badge. They wouldn't let the kid in. She was furious! I didn't see her angry that often, but I saw her angry that day."[72]

Meanwhile, Millie continued to spend quite a bit of time over at MIT. Among other things, in 1965 her former boss, Ben Lax, had resigned his post at Lincoln so he could work full time at the National Magnet Laboratory, which he led in the effort to establish and then direct on the MIT campus in Cambridge. Following the lab's opening, Millie took advantage of the new, high-intensity magnets and strong magnetic fields they could create by designing and running numerous experiments on graphite and other materials. She had several unofficial graduate students during these years, starting with Samuel Williamson, who would become a successful biophysicist and pioneer in using magnetic fields to measure activity in the brain.[73]

At MIT, Millie also found a little more freedom to be both a working scientist and a mom to young children. In an oral history published in 2020, Ben Lax, who died in 2015, recalled, "One day, while Millie was running an experiment, she called me in to show me something, and in one of the spare rooms there was one of her little babies. Apparently they made a nursery where, in between magnet runs, she would nurse the baby. . . . I thought [it] very appropriate for the kind of spirit that prevailed in the Magnet Lab, a very free spirit, a very cooperative spirit."[74]

In the mid-1960s, an administrative change at Lincoln Laboratory presented a huge question mark as to Millie's ability to continue as a full-time researcher and mother to four young children. As Millie explained it, all staff scientists at the time were notified that they had to start clocking in promptly at 8:00 a.m. each day.[75] "It was very hard when the children were small to be at the lab at 8 o'clock," Millie said in a 2002 interview.[76] To be sure, her actual work was never criticized, and she put in plenty of time in working from home in the evenings. Some of the time, Gene would drive her to the lab, so they'd both be late, but only Millie would get hassled about it.[77] "My supervisor at Lincoln Lab complained about me so much that I got tired of hearing all of the complaints, because I was doing the best that was humanly possible," she said, adding pointedly, "The people who were judging me were all bachelors."[78]

Millie's colleague Laura Roth had faced the same problem. She eventually left Lincoln for Tufts University, and Millie began quietly letting friends and sympathetic colleagues know that she felt stuck. In addition to the time dispute, she also found it had become harder to secure federal funding for basic research than it had been during the post-Sputnik era of the early 1960s, and new accountability rules, forcing researchers to justify their work in rather tedious ways, began to stifle creativity.[79] "I let people know that times were really tough," she admitted in 2009. "I didn't know how I was going to continue my career."[80]

One of Millie's friends, a former Lincoln colleague who had recently transitioned to the MIT faculty, wanted to help. George Pratt, a professor of electrical engineering, had previously invited her to guest-lecture in a number of his classes, and he knew she was both a strong scientist and educator.[81] "He had pity on me," Millie explained at a party in 2007, "and he said, 'This rule is ridiculous when you're so productive. Why do you have to start at 8 o'clock? Maybe they can make an exception for you while your children are so small.' But they didn't make an exception."[82]

Over lunch one afternoon while Millie was working on campus, Pratt floated the idea that she'd make an excellent visiting professor in the Department of Electrical Engineering. Would she be interested? "Sure!" she blurted out with a smile. "If you offer me a job, I'll come and spend a year here."[83]

Pratt had heard about a unique MIT endowment that supported visiting faculty who were women and thought it could be perfect for Millie. It was, as historian of science Margaret Rossiter described in her three-volume *Women Scientists of America*, one of several "coercive philanthropy" efforts underway at the time, at institutions such as Brown University, Harvard University, and the University of Michigan, that were meant to increase the presence and representation of women on college faculty around the country. MIT's endowment, initiated by and named for Abby Rockefeller Mauzé, the eldest of the six children of philanthropists John D. Rockefeller Jr. and Abigail G. Aldrich, was established in 1963 with a $400,000 gift from Laurence Rockefeller and the Rockefeller Brothers' Fund. The first Abby Rockefeller Mauzé Professor was distinguished British X-ray crystallographer and Nobel laureate Dorothy Hodgkin, who spent a week at MIT in 1965, giving lectures on her work and meeting with students and faculty.[84]

With a potential funding source all but waiting to be tapped, George Pratt and Louis Smullin, then head of the MIT Department of Electrical Engineering, nominated Millie for a one-year professorship at the Institute. Millie, meanwhile, applied for and quickly received a grant from the Rockefeller Mauzé endowment. While she would continue to work on special projects at Lincoln Laboratory for about five more years, once she accepted a position as a visiting professor, MIT became Millie's professional home for the rest of her illustrious career—and, indeed, for the rest of her life.[85]

In the end, Millie, with Gene's immeasurable help, had spent seven outstanding years at Lincoln Laboratory, quantifying the magneto-optical properties of graphite and a host of other semimetals. Gene stayed on at Lincoln until the mid-1970s, when he rejoined Millie at MIT. In the interim, they continued to collaborate on a number of research projects—and, of course, on being doting parents to Marianne, Carl, Paul, and Eliot.[86]

6 MENS ET MANUS

The *Boston Globe*, October 8, 1967: Set between a studio portrait of five tiara-wearing beauty pageant winners scheduled to march in the Columbus Day Parade and a tongue-in-cheek article on students committing "minor indiscretions" related to a World Series appearance by the Boston Red Sox, a beaming visage of thirty-seven-year-old Millie Dresselhaus trumpets her arrival at one of the world's leading institutions of higher learning. "Dr. Mildred S. Dresselhaus, of Lincoln Laboratory, who has achieved prominence as a solid state physicist," the announcement stated, "has been appointed Abby Rockefeller Mauzé Visiting Professor at the Massachusetts Institute of Technology."[1]

When Millie agreed to join the MIT faculty in fall 1967, it was no ordinary job acceptance. Though she was a visiting professor, she made history as the first female professor within the MIT Department of Electrical Engineering and the second within the entire MIT School of Engineering.[2]

It may have seemed an unusual placement at the time: she had no formal engineering background. During the 1960s, however, the MIT physics department was concentrated heavily on particle physics (that is, high-energy physics), the subfield that focuses on the behavior of individual atoms and their constituent parts rather than the bulk properties of material substances. Moreover, the understanding of materials was becoming ever more critical to the engineering of electronic devices, so the Department of Electrical Engineering leadership wanted to be sure their students were being properly supported with the most appropriate science education.[3] "This was the heyday of semiconductor research, and the industry was growing rapidly," Millie noted in 2012. "No one could anticipate the sort of skills our students would need to invent the next generation of semiconductor devices, so a broad background . . . was very much desired."[4]

For her appointment, one of Millie's primary roles was teaching physics to MIT's standout engineering students. According to historian of science Joseph Martin, this assignment filled an urgent need at MIT, where solid-state physics research was strong but courses in the rapidly evolving field were limited. "Dresselhaus's willingness to present the latest in solid state theory in a way that emphasized its utility for practically minded physicist and engineers addressed the considerable appetite for such a course among MIT students," Martin wrote in 2019.[5]

The task felt like second nature to Millie. In addition to her well-established teaching credentials, her prior research had given her valuable experience in interdisciplinary settings. This was true at both the University of Chicago—where, in the absence of a proper adviser, she'd rounded up individuals from very different backgrounds to help her design experiments—and Lincoln Laboratory, where she'd collaborated with scientists in various fields. "I was very comfortable with talking to people, explaining physics principles to them," she said in 2001.[6] Her early efforts to develop curricula that would appeal to students in various fields was immediately successful; in fact, two of the courses she developed early on at MIT—on the physics of solids and of solid-state applications—are still taught at MIT today.[7]

"She taught by handing us clear and complete handwritten class notes and asking us to listen, understand, and participate actively by asking questions, just as Fermi had taught her at Chicago," wrote physicist Aviva Brecher, an early student who took Millie's first solid-state physics course at MIT, in a 2017 remembrance. "She was the best, clearest, and most caring teacher we had."[8]

Beyond the added teaching, Millie's responsibilities were mostly unchanged from her time at Lincoln Laboratory. Her research continued largely uninterrupted, as she'd already been doing quite a bit of it at the Magnet Lab facility on campus. She had previously been advising MIT students while at Lincoln; now she simply had more of them under her wing and in a more formal capacity. She even continued to work at Lincoln a day or so each week.[9]

One new activity, however, was unlike anything she'd taken on before. Millie was quite aware of the fact that her opportunity to serve as a visiting professor came directly from Abby Rockefeller Mauzé, whose vision when she and her brothers established a chair at the Institute was to encourage and support women in academia. Millie felt she needed to give something back to support the mission of the chair's founding. And so, beginning in fall 1967, she turned to Emily Wick, a food scientist and the associate dean of students who was the first woman to earn tenure at MIT. With backing from Wick, Millie decided to spend an hour or two every week meeting with young women of MIT and providing encouragement, advice, and a sounding board for their frustrations.[10]

At this time in MIT's history, only 5 percent of the undergraduate student body was female (today it's about 47 percent), and the percentage of female faculty was even lower. Nationwide, women in the 1960s and 1970s were vastly underrepresented in science and, especially, engineering. Among other things, women had only just begun taking advantage of new laws such as the Civil Rights Act of 1964, which rendered illegal workplace discrimination based on race, sex, religion, or national origin, and Title IX of the Education Amendments of 1972, which provided new protections from sex-based discrimination in education programs or activities receiving federal funding.[11] "Our women students needed role models, so I tried to help," Millie stated in 2002.[12]

It seemed obvious to Millie that if female students were having trouble, it was largely because they were so drastically outnumbered by male peers. She wasn't at first sure what she might do to alleviate this problem, but her initial months on the faculty provided a particularly eye-opening glimpse into the situation and would give her ideas for future directions in which to implement positive changes. As we will see, Millie was destined to support women and other underrepresented students in critical ways for the rest of her career—at MIT and elsewhere.[13]

It didn't take long for Millie's colleagues to realize she was an invaluable asset to the electrical engineering department and to the Institute as a whole. Shortly after arriving as a visiting professor, she received an offer of a permanent full professorship with tenure. "That was amazing!" she recalled in a 2009 interview.[14]

It's hard to overstate how unusual such a professional leap would be today. Even the most talented and promising young professors must navigate the steps of the tenure track, including several key reviews and promotions, before they're eligible for full professorship with tenure. It also bears noting that in the late 1960s, many prestigious colleges around the United States did not even accept women as regular students. Yale, Princeton, Duke, Brown, Johns Hopkins, and Harvard—none of these standout universities allowed women to apply for regular admission in 1968, when Millie permanently joined the MIT faculty.[15] MIT had allowed a small number of women to matriculate for nearly one hundred years to that point, but hiring a female engineering professor at the full professor rank and granting her tenure without needing to jump over the usual hurdles was extraordinary.

Millie's colleagues clearly wanted her on their team, and they were willing to make a big bet on her future in order to entice her to stay at MIT. She later credited Louis Smullin, head of the electrical engineering (EE) department, for seeing her potential, despite the fact that she was working in an area in which very few people were interested at that time—and for providing her with an enormous opportunity to change the way the world thinks about carbon. But it was the students she'd supported who were the real springboards for Millie's transition.[16] "Millie was a superb teacher. . . . She taught the best physics course I ever took at MIT and raised the bar for teaching," said former student Aviva Brecher in 2017. "We were so impressed with her teaching skill . . . that we petitioned Lou Smullin, then head of EE, to extend her tenure from a visiting professorship, which he did."[17]

Millie promptly accepted the offer. In so doing, she once again made history, this time as the first tenured woman in the MIT School of Engineering and one of the first two women to achieve the rank of full, tenured professor in all of MIT (the other was Emily Wick, who was promoted to full professor on the same day Millie's full professorship began). It was a strong match that would continue for five decades. True to MIT's motto, *mens et manus*—for mind and hand—the Institute provided Millie a professional home where she could engage her well-established scientific instincts and make new inquiries into engineering applications.[18]

Her appointment also gave her newfound confidence that she was doing what she was really meant to do. Some years after joining the faculty, Millie revealed publicly that despite her self-assured veneer, she'd doubted on numerous occasions whether she would be able to continue her research journey, due mostly to her status as a woman at a time when female scientists and engineers were still a rarity. Of course, Rosalyn Yalow, Enrico Fermi, and Gene Dresselhaus, among others, had given her immeasurable support and encouragement along the way, but it is words and actions of *dis*couragement that often fester, inviting anxiety and doubt. Although he later apologized and all was forgiven, the dismissive attitude of Millie's PhD supervisor during her graduate school years stuck with her for quite some time. "I trusted his judgment," she said to *Cosmopolitan* magazine in a 1976 profile. As a result of his philosophy on women in science, she admitted, "I had very low expectations for myself. It wasn't until I became a full professor in the prestigious MIT electrical engineering department that I began to take my career seriously."[19]

Thankfully, plenty of others were already taking her seriously. One of her greatest champions on joining the MIT faculty was a fellow MIT physicist, some three decades her senior, Arthur von Hippel. The two had first met when von Hippel's string quartet, organized through the MIT electrical engineering department, was recruiting a violist—and Millie answered the call. A German-born scientist who fled the Nazis and came to lead MIT materials science relating to the development of radar, von Hippel would become a lifelong mentor and friend. He was fascinated by Millie's work and provided support for her foray into the electronic properties of materials, serving too as an example in numerous settings, from advocating interdisciplinarity to encouraging students and collaborators from all levels and backgrounds.[20] "Professor von Hippel was always telling me that new and interesting high-impact ideas are most likely to come from adventurous young people," Millie wrote in a 2014 remembrance. "His appreciation of these revolutionary and innovative ideas was one of his long-term impacts on me and others in my age group."[21]

And so, with an official transfer to MIT, Millie found herself in a comfortable place with a strong research focus, support from funding agencies and colleagues, happy and energetic students, and the flexibility to work where and when she needed in order to support her family. With the question of her future in science all but answered, it was time to get to work.

A CRITICAL ABOUT-FACE

For anyone with a research career as long and as accomplished as that of Mildred S. Dresselhaus, there are bound to be certain papers that might get a bit lost in the corridors of the mind—papers that make only moderate strides, perhaps, or that involve relatively little effort or input (when, for example, being a minor consulting author on a paper with many coauthors). Conversely, there are always standout papers that one can never forget—for their scientific impact, for coinciding with particularly memorable periods of one's career, or for simply being unique or beastly experiments.

Millie's first major research publication after becoming a permanent member of the MIT faculty fell into the standout category. It was one she described time and again in recollections of her career, noting it as "an interesting story for history of science."[22]

The story begins with a collaboration between Millie and Iranian American physicist Ali Javan. Born in Iran to Azerbaijani parents, Javan was a talented scientist and award-winning engineer who had become well known for his invention of the gas laser. His helium-neon laser, coinvented with William Bennett Jr. when both were at Bell Labs, was an advance that made possible many of the late twentieth century's most important technologies—from CD and DVD players to bar-code scanning systems to modern fiber optics.[23]

After publishing a couple of papers describing her early magneto-optics research on the electronic structure of graphite, Millie was looking to delve even deeper, and Javan wanted to help. The two met during Millie's work at Lincoln Lab; she was a huge fan, once calling him "a genius" and "an extremely creative and brilliant scientist."[24]

For her new work, Millie aimed to study the magnetic energy levels in graphite's valence and conduction bands. To do this, she, Javan, and a graduate student, Paul Schroeder, employed a neon gas laser, which would provide a sharp point of light to probe their graphite samples. The laser had to be built especially for the experiment, and it took years for the fruits of their labor to mature; indeed, Millie moved from Lincoln to MIT in the middle of the work.[25]

If the experiment had yielded only humdrum results, in line with everything the team had already known, it still would have been a path-breaking exercise because it was one of the first in which scientists used a laser to study the behavior of electrons in a magnetic field. But the results were not humdrum at all.[26] Three years after Millie and her collaborators began their experiment, they discovered their data were telling them something that seemed impossible: the energy level spacing within graphite's valence and conduction bands were totally off from what they expected. As Millie explained to a rapt audience at MIT two decades later, this meant that "the band structure that everybody had been using up til that point could certainly not be right, and had to be turned upside down."[27]

In other words, Millie and her colleagues were about to overturn a well-established scientific rule—one of the more exciting and important types of scientific discoveries one can make. Just like the landmark 1957 publication led by Chien-Shiung Wu, who overturned a long-accepted particle physics concept known as conservation of parity, upending established science requires a high degree of precision—and confidence in one's results. Millie and her team had both.[28]

What their data suggested was that the previously accepted placement of entities known as charge carriers within graphite's electronic structure was actually backward. Charge carriers, which allow energy to flow through a conducting material such as graphite, are essentially just what their name suggests: something that can carry an electric charge. They are also critical for the functioning of electronic devices powered by a flow of energy.[29]

Electrons are a well-known charge carrier; these subatomic bits carry a negative charge as they move around. Another type of charge carrier can be seen when an electron moves from one atom to another within a crystal lattice, creating something of an empty space that also carries a charge—one that's equal in magnitude to the electron but opposite in charge. In what is essentially a lack of electrons, these positive charge carriers are known as holes (figure 6.1).[30]

Millie, Javan, and Schroeder discovered that scientists were using the wrong assignment of holes and electrons within the previously accepted structure of graphite: they found electrons where holes should be and vice versa. "This was pretty crazy," Millie stated in a 2001 oral history interview. "We found that everything that had been done on the electronic structure of graphite up until that point was reversed."[31]

As with many other discoveries overturning conventional wisdom, acceptance of the revelation was not immediate. First, the journal to which Millie and her collaborators submitted their paper originally refused to publish it. In retelling the story, Millie often noted that one of the referees, her friend and colleague Joel McClure, privately revealed himself as a reviewer in hopes of convincing her that she was embarrassingly off-base. "He said," Millie recalled in a 2001 interview, "'Millie, you don't want to publish this. We know where the

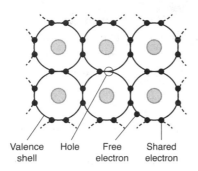

Valence   Hole   Free   Shared
shell   electron   electron

FIGURE 6.1
In this simplified diagram, electrons (black dots) surround atomic
nuclei in a crystal lattice. In some circumstances, electrons can break
free from the lattice, leaving an empty spot or hole with a positive
charge. Both electrons and holes can move about, affecting electrical conduction within the material.

electrons and holes are; how could you say that they're back-
wards?'"[32] But like all good scientists, Millie and her colleagues
had checked and rechecked their results numerous times and
were confident in their accuracy. And so, Millie thanked
McClure and told him they were convinced they were right.
"We wanted to publish, and we . . . would take the risk of ruin-
ing our careers," Millie recounted in 1987.[33]

Giving their colleagues the benefit of the doubt, McClure
and the other peer reviewers approved publication of the paper
despite conclusions that flew in the face of graphite's established structure. Then a funny thing happened: bolstered by
seeing these conclusions in print, other researchers emerged
with previously collected data that made sense only in light
of a reversed assignment of electrons and holes. "There was a
whole flood of publications that supported our discovery that
couldn't be explained before," Millie said in 2001.[34]

Today, those who study the electronic structure of graphite do so with the understanding of charge carrier placement

gleaned by Millie, Ali Javan, and Paul Schroeder (who ended up with quite a remarkable thesis based on the group's results). For Millie, who published the work in her first year on the MIT faculty, the experiment quickly solidified her standing as an exceptional Institute researcher. While many of her most noteworthy contributions to science were yet to come, this early discovery was one she would remain proud of for the rest of her life.[35]

ADVANCING MIT WOMEN

From its creation in 1884, MIT's Margaret Cheney Room has served as a special sanctuary—a space within the Institute that's designated for women. It predates the Institute's existing Cambridge campus and was originally located at the intersection of Boylston and Clarendon Streets on the original MIT campus in the Copley Square area of Boston—a busy part of the city that today is taken up mostly by office buildings and, notably, the John Hancock Tower, the tallest structure in New England. The room is named for Margaret Swan Cheney, a member of the MIT Class of 1882 and a disciple of the pioneering chemist Ellen Swallow Richards, MIT's fearless first female student and instructor, who served as a champion for women during the Institute's earliest years. Cheney was a passionate student of science who died in 1882 after a brief illness. Following her death, it was proposed that a women's reception room in the works for MIT's new Walker Building be named in her honor. The Cheney family was delighted by this and donated $500 to help establish the first Margaret Cheney Reading Room. When MIT relocated across the Charles River to Cambridge in 1916, the Cheney Room moved along with it. Today, it comprises several small rooms overlooking the Institute's Killian Court and serves as a haven for self-identified women and nonbinary and gender-nonconforming individuals.[36]

Starting in 1968, in the safety of the Cheney Room, women from around the Institute began to gather every two weeks for tea, crumpets, and conversation. Some came to vent. Others simply wanted to hear what their sisters in science and engineering were up to that month and to count themselves as part of a movement: an influx of women in the STEM fields at MIT and around the United States during the heady 1960s and 1970s. With the civil rights and women's movements playing out on both local and national stages, it was a time of great change—and much of it would ultimately make academia a more inclusive universe. But the process of making change can often be rocky, and MIT's female students were among those who felt the growing pains.[37]

When Millie and Emily Wick decided to hold regular gatherings for female students that year, being a woman in academia was still seen as something of an aberration, especially at a place like MIT, where women often faced a classic catch-22. In student newspaper cartoons, they were criticized for being at once too feminine—unable to hack it in a highly technical and often competitive environment—and not feminine enough—frumpy, boring, and obsessed with academics. To be sure, young women who managed to secure a spot at any institution of higher learning then didn't even garner the respect of being called "women" or "students"; they were "coeds," a loaded term that effectively pointed a finger at women for breaking up a sphere where only men had existed previously.[38]

As Millie herself experienced firsthand during her graduate and postdoctoral years, being among the first to take up space where men had formerly reigned could be a tricky thing to negotiate, in part because sexist or otherwise biased attitudes were not uncommon among male students and faculty during this period. Indeed, although MIT was one of the first technical colleges in the United States to admit women, the

Institute's policies toward female students through the first half of the twentieth century were haphazard at best. During the 1960s and 1970s, as the percentage of female students grew from single digits to high teens, the Institute still struggled with how (and whether) to recruit female students—and how to support them once they matriculated.[39]

"MIT was in need of a platform to develop and build opportunities for women students and faculty to pursue successful careers in science and engineering," said Sheila Widnall, longtime MIT professor of aeronautics and astronautics who was one the Institute's few female faculty members in the early 1960s, in a 2017 remembrance. "Millie was crucial in the building of that platform."[40] And yet, although she had plenty of teaching under her belt, Millie was fairly green when it came to mentoring and counseling. As a result, her early efforts to support budding scientists and engineers served as a learning experience, broadening her own view of the status of women in the STEM fields.[41] Of the students she saw at those first gatherings, Millie noted to *Science* in 2014, "One thing they mentioned was how isolated they felt . . . they wondered if they belonged there."[42]

In her 1976 MIT oral history, Millie described in greater detail some of the discussions that took place during her Cheney Room gatherings—including many students' admission that they struggled in the face of biased attitudes and a prevailing sense of exclusion: "The undergraduates had a lot of problems with the male students; that was a perennial topic. They were so outnumbered in the technical courses. . . . Many felt that they weren't performing up to their potential because they felt so self-conscious. And the professors didn't know how to deal with them. . . . Having me around, having the other women students around, and having the opportunity to talk about these things made a big difference to [attendees], and sometimes made the difference between quitting or staying at MIT."[43]

Carol Steiner, a professor of chemical engineering at the City College of New York and MIT alumna from the class of 1976, echoed these sentiments at a 2017 tribute event: "As an undergraduate . . . we had a 10 to 1 male-to-female ratio. Just knowing that Millie was there provided some inspiration and encouragement."[44]

Women of color could have an especially difficult time feeling that they belonged in higher education—and particularly at a technical institute such as MIT—as they faced a classic double bind in which both their gender and their race seemed to work against them.[45]

"When I entered MIT as an undergraduate, I was one of just two African-American women in my class," Shirley Ann Jackson, president of Rensselaer Polytechnic Institute and the first African American woman to earn a doctorate from MIT, stated in a 2017 remembrance. "The other students were often unwelcoming to me, and some of the professors could be equally so. At one point, thinking of majoring in physics, I asked a distinguished professor for advice. His response was, 'Colored girls should learn a trade.' Now, the advice offered by Millie Dresselhaus, needless to say, was entirely different. She had a much more panoramic view of life, and MIT was fortunate to have her. . . . Millie's patience as a teacher, her unwillingness to allow struggling students to quit, and her efforts to break down the institutional barriers for young women in science, including me, were a call to action for all of us who followed her and who seek also to open up opportunities for young people from every possible background and origin."[46]

Millie and Emily Wick, along with Wick's assistant, Dottie Bowe, set about brainstorming concrete ways to help MIT's female students feel validated. The three were among the few women in high administrative or other leadership roles at the time, and they felt a responsibility to support undergraduate and graduate students however they could. Wick was widely known as a champion of women on campus; in her role as associate dean of students, she focused much of her energy on supporting women.[47] "At that time, [Emily Wick] was doing practically everything that was being done for women students at MIT," Millie stated in a 1976 oral history. This included managing various programs for women as well as reading through applications from all of the high school girls applying to MIT.[48]

Within her first year at MIT, Millie and Wick decided to familiarize themselves with the process by which women were admitted as undergraduates. Although MIT had been admitting female students here and there since 1870, it wasn't until the early 1960s, which saw the opening of a women's dormitory, that the MIT administration began to pay sustained attention to recruiting and supporting women. Wick invited Millie to share in the reading of materials submitted from female applicants—about four hundred per year by the late 1960s. She readily agreed, despite the significant addition to her already packed workload and family life these readings would entail.[49]

Paying close attention to the applications was a fruitful exercise that brought to light several key concerns that would lead to important changes in MIT's admissions process. "It was believed at that time that admitting women was somewhat risky, because they weren't doing well," Millie stated in a 2007 interview. "But upon looking at the situation, I felt that it wasn't the women, but it was the environment that they were in that contributed to their sub-critical performance."[50]

Adjusting MIT's climate to be more welcoming of women— especially women of color—would be an ongoing project. At the time, though, Millie believed that making sure more women were accepted in the first place would be the biggest way to create positive change. The most substantial admissions insight that she and Wick uncovered was that male and female applicants were being considered for admission based on entirely different criteria. As a result of very few housing options for women—a limiting factor that directly determined how many female students the Institute accepted—there was more competition among female applicants. This meant that in order to be accepted, women's test scores had to be higher than men's, and their recommendations had to be stronger too. Millie stated plainly in 1976, "it was harder for women to get into MIT than for men."[51]

Based on their research, Emily Wick submitted an official report making a number of recommendations to MIT's admissions committee. At the top of the list was removing the separate and unequal admissions process for women so that all applicants were considered on essentially the same grounds. The report also advocated for more equal recruiting of talented high school students—women and men—as well as the development of better outreach materials that would showcase the fact that MIT accepted and supported women. It further argued that accepting more female students would serve to improve women's performance in areas in which they had been struggling, and it recommended a system of evaluating applications that included faculty input.[52]

Many of their recommendations were adopted in one form or another. The result—dovetailing on nationwide legal and social pressures for increased acceptance of women in the workforce—was an immediate and marked uptick in the percentage of female students at MIT (figure 6.2).[53]

A FORUM FOR WOMEN

In 1971, Emily Wick stepped down from her position as associate dean of students to return to her professorial duties.

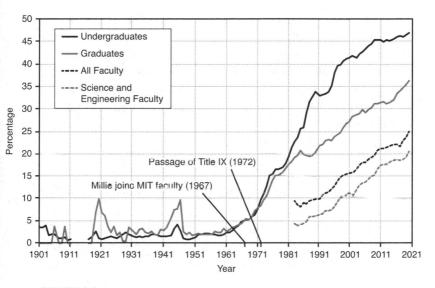

FIGURE 6.2

Female students and faculty at MIT, 1901–2020. Faculty data are not available prior to 1983. Credit: Data provided by the MIT Office of the Provost/Institutional Research.

The decision on the part of the MIT administration not to replace her did not sit well with many female students, who had come to rely on Wick as an important ally within the Institute administration. So in January 1972, shortly after Wick officially left her post, she, Millie, and Dottie Bowe organized an open session during the Institute's independent activities period to discuss issues that were facing MIT's undergraduate women. When the day arrived, the trio were stunned to find one hundred or so people from all walks of the MIT community—mostly women plus a couple of men—crammed into the Cheney Room, each hoping to share problems, concerns, and suggestions for the future of women at the Institute.[54]

Thus was born the MIT Women's Forum, a regular gathering of individuals from across campus that provided a space in which to examine the plight of MIT women. Over the following weeks, attendees aimed to convince the administration to systematically review policies and inequities felt by women at the Institute, relating to everything from child care to academics to financial aid. The administration listened, and a committee chaired by Millie and senior MIT student Paula Stone was formed to look carefully at these and other concerns.[55]

Around this same time, Millie was appointed associate head of the Department of Electrical Engineering under department head Louis Smullin. This significant leadership portended a big change: not only was Millie the only female faculty member in MIT's largest department, she was also the first woman in all of MIT to take on such a high-level academic post. The Associated Press was moved to write a brief article about Millie's new position in February 1972, which meant that her promotion was announced in newspapers around the nation.[56]

Besides its historic nature, Millie's appointment was about to add many new responsibilities to her already-packed schedule; this prompted her to formally end her continued relationship with Lincoln Laboratory. Not insignificantly, it also provided Millie with a bit of additional gravitas with which to approach her work on the Women's Forum—which, notably, coincided with the birth of Title IX, the US civil rights law passed in June 1972 that prohibits sex discrimination in education programs operating through federal funding.[57]

Owing to the varied nature of its participants—and the disparate tasks these subgroups aimed to further around the Institute—the Women's Forum would take on assorted projects, such as creating a women's kiosk on campus and organizing a talk by women's rights leader Gloria Steinem. It also produced at least two influential reports on the status of women at MIT.[58] For these, Millie received pivotal advice from Jerome Wiesner, MIT's president at the time, who was generally supportive of the forum but also critical of an early draft of its first report. "He was my first and severest mentor on science policy," Millie said in a 1997 roundtable on the history of women at the Institute. "What he said was you have to treat science policy and service at MIT and elsewhere—the country—in the same way you treat your own research . . . you have to fashion any proposal and document your issue with the same kind of excellence as is expected of your MIT scholarly work. That stayed with me, it was terrific advice."[59]

The Ad Hoc Committee on the Role of Women Students at MIT ultimately described in its 1972 report to the MIT administration a wide spectrum of dispiriting attitudes and behaviors toward female students, from bias and microaggression to downright enmity.[60] Chaired by Millie and Paula Stone, the committee noted, "It does appear appropriate to acknowledge at the outset, that there are many positive aspects of life at MIT for women students and that many of us have been and are happy here. But it is equally clear that attitudes and policies should be and can be improved significantly."[61] The report went on to list eleven areas that the authoring committee deemed ripe for improvement in terms of how female students were being treated at the time, from admissions and the overall academic environment to student activities, housing, and medical treatment.

"A discriminatory attitude against women is so institutionalized in American universities as to be out of the awareness of many of those contributing to it," the report stated, adding:

> If many people (professors, staff, male students) at the Institute persist in feeling that women jeopardize the quality of MIT's education, that women do not belong in traditionally male engineering and management fields, that women cannot be expected to

make serious commitments to scientific pursuits, that women lack academic motivation, that women can only serve as distractions in the classroom, then MIT will never, and can never, be a coed institution with equal opportunities for all of its members. . . . Although we realize that the majority of the discrimination is not intentional, it nevertheless does exist. The outcome is that some female students at MIT feel that they are not being taken seriously, that they are not being treated on the same basis as the male students, and that they are being educationally shortchanged.[62]

As historian Amy Sue Bix noted in her book, *Girls Coming to Tech!*, the report "represented a self-directed rallying cry, telling MIT women that gender discrimination would change only when female students, faculty, and staff organized to demand improvement."[63] The authors acknowledged that some of the report's recommendations were not likely to be addressed in short order, but it nevertheless made big waves at MIT and peer institutions. One immediate outcome was that the administration appointed Mary Rowe as a new high-level advocate for female students. Initially a special assistant to the president and chancellor for women and work, Rowe eventually became an ombuds, providing neutral, confidential advice and conflict management help to anyone—regardless of gender—who came through her door. (A similar position, with the charge of supporting students of color at MIT, was created about a year after Rowe began. Special assistant to the president and chancellor for minority affairs Clarence Williams would go on to serve with Rowe as an ombuds and an adjunct professor of urban studies and planning.)[64]

As a result of her efforts to support women at MIT, Millie herself began to think differently about equality in education and in the workplace, in the STEM fields in particular. Although she did not identify with the type of radical feminism that was gaining momentum in the United States at the time, she couldn't help but be moved by the experiences of so many women with legitimate complaints about biased or discriminatory treatment. She began to recognize that she benefited immensely from inner confidence, a will to persevere in the face of obstacles, and the backing of influential men in her personal and professional spheres. She internalized that in the face of discriminatory attitudes, some very capable women and other underrepresented students suffered due to a lack of such critical supports.[65]

"I saw her change her opinions as she began to recall incidents . . . that she'd previously managed to ignore," wrote University at Buffalo professor of physiology Susan Udin, an MIT alumna and former Women's Forum participant, in 2017. "I think that that sort of blindness had served a useful purpose for her in the first part of her career."[66]

"Up to the time of the Ad Hoc Committee report and the [Women's] Forum, I always figured that women could take care of themselves," Millie admitted in the late 1970s. "I overlooked, in a way, all that help that I got along the way. I wasn't really sensitive that women had to help other women. . . . Now it was up to me to help the others."[67]

GLOBAL IMPACT

Summertime in the early 1970s provided Millie, Gene, and their children with opportunities to visit other countries for a mix of family vacation, professional networking, and extracurricular teaching. These working holidays allowed Millie and Gene to connect with international colleagues, some of whom had MIT ties, and to forge new physics connections in distant parts of the world.

The first of these extended family outings was a 1971 sojourn to Brazil. Millie and Gene were among a number of experts invited to teach intensive courses to professors there as a way to jump-start solid-state physics research across the nation. Over time, this and subsequent visits helped to establish Brazil as a leading contributor to research in this field.[68]

As part of their visiting professorships, the Dresselhauses made sure that their children were included and that they had plenty of time to relax and explore with Marianne, Carl, Paul, and Eliot.[69] "Millie had no problem taking four small children . . . halfway around the world," her son, Paul, said at a tribute event in 2017. "We were not well-behaved children, so I'm not exactly sure how she did it. But . . . she just did it. She didn't accept the impossibility of traveling."[70]

While Gene and Millie taught courses by day, their children—at that time seven to eleven years old—kept themselves amused with any number of local activities. Although a babysitter nominally watched them when they stayed at their accommodations on a large farm, Millie and Gene gave the children the freedom to explore zoos and other places of interest on their own, with the eldest, Marianne, in charge.[71] "Marianne was like the mother; she had to take care of her brothers," says Eliot, one of the brothers.[72]

In 1972, the summer after a busy semester of teaching, research, and her Women's Forum activities, Millie traveled with Gene and the children to Israel, where she had a visiting professorship at the Israel Institute of Technology, also known as Technion. In the years that followed, the Dresselhauses enjoyed similar visits to Japan and Venezuela, all of which advanced important research collaborations.[73] "Being around Millie and her students, postdocs, and collaborators was always an international experience," says Paul Dresselhaus.[74]

These trips, and the family and collegial bonding they facilitated, also served as well-timed breaks, as the mid-1970s would prove to be highly consequential and extremely busy years for Millie and Gene back in Massachusetts. In addition to parenting their four growing children, Millie continued both her academic leadership and her advocacy for women across MIT with new projects that required an intense amount of work, and her research advanced at breakneck speed: she and Gene worked together to help launch a new field of condensed-matter physics—one that kept them busy for nearly twenty years and would lead to fundamentally new understandings of carbon as a useful material.[75]

7 WELCOME TO THE NANOWORLD

It was the year of *Roe v. Wade*, the Watergate scandal, and the end of US involvement in the Vietnam War. The first handheld cell phone call provided a harbinger of our soon-to-be always-on world; Billie Jean King beat Bobby Riggs in tennis's famous "battle of the sexes"; and hip-hop began to emerge as a unique musical genre, thanks to DJ Kool Herc and others. The year was 1973, and Millie Dresselhaus was just getting warmed up.[1]

Leaves on the MIT sycamores were barely beginning to brown when Millie made the decision to step down from her leadership role on the MIT Women's Forum after a year and a half of sustained contributions. Rest assured, however, her advocacy work wasn't going anywhere; it was just shifting.[2]

In fact, Millie was honored, almost simultaneously, with a prestigious fellowship from the Carnegie Foundation that focused on supporting women at a handful of universities around the United States. Wasting no time, she recruited fellow MIT engineering professor Sheila Widnall to help her translate the opportunity into the creation of an entirely new kind of course for first- and second-year undergraduates. Titled "What Is Engineering?" the course's goal was to provide an academic offering that would help women catch up to their male peers, who were far more likely to begin their college careers with previous exposure to engineering subjects and projects from childhood and adolescence, either formally in school or informally through relatives, friends, or others. Indeed, well before the term *gender gap* even existed, the course sought to level the playing field by opening the challenges and opportunities of engineering to everyone.[3]

"Weekly classes presented information about careers in engineering, including many presentations from working engineers in the Boston area," Widnall noted in a 2017 remembrance. The class also, according to Widnall, featured labs providing students with hands-on experiences in electronics, welding, and model building. "We had hoped for 15 students per semester, but we got over 100, half of whom were men. Many MIT women and minority students took the course, and quite a few decided to major in engineering."[4]

"The reason we got into this course was to acquaint women with what was happening in the engineering school and telling them there was a place for them in engineering," Millie explained to Clarence G. Williams, adjunct professor emeritus of urban studies and planning at MIT, for his book, *Technology and the Dream: Reflections on the Black Experience at MIT, 1941–1999*. "I think that this course also had some significant impact on getting minority students to take majors in the engineering school," she added, "or at least to take some fraction of a minor in engineering."[5]

Around this same time, Millie was honored with an appointment to a new Abby Rockefeller Mauzé endowed chair at MIT. This semipermanent chair was established through the same endowment as her 1967 visiting professorship, and Millie was the first individual to hold it as a nonvisiting professor.[6]

In yet another move to support women at MIT, Millie again teamed up with Sheila Widnall to organize the first Women's Faculty Luncheons, initially with money from her Carnegie

Foundation fellowship and later from the Mauzé chair.[7] She and Widnall took it upon themselves to mentor the few female faculty members at MIT at the time "because they were in all kinds of different departments, and many [of those departments] had no women," Millie recalled in recalled in 2007. "They were the first women faculty, and some of the men felt uncomfortable in telling them what they should be doing to promote their careers."[8]

"We discussed the tenure process, how to get grant money, the mentoring of women students, and other inequities and how to fix them," Widnall added in a 2017 remembrance. "We had broad support from the senior administration of MIT and, typically, we would invite a dean or a provost to attend these lunches and speak with the women faculty."[9]

CARBON CONCOCTIONS

While she was committed to doing whatever she could to make life easier for underrepresented students and faculty around the Institute, Millie in the early 1970s was spending even more time developing a new research venture that would continue for nearly two decades and would ultimately lead to some of her greatest contributions to science.[10]

You'll recall that graphite is composed of slippery flakes: single-atom-thick sheets of carbon—graphene—that combine together to form a kind of carbon confection. In 1973, Millie began experimenting with a new recipe: graphite embedded with bits and pieces of other elements. This layering of graphene sheets with external additives is known as intercalation—and the resulting mix of carbon layers with so-called guest species is called a graphite intercalation compound. These synthetic compounds come in a myriad of flavors. With a multitude of atoms and molecules that can squeeze in between carbon layers, a visit to the graphite bakery would feature hundreds of intercalation varieties, displaying properties that diverge quite substantially from those of pure graphite.[11]

It's important to note that graphene was known to scientists well before Millie and Gene began working on it. A 1947 paper by Philip Wallace, for example, first laid out an early band structure and served as a blueprint for future insights into the carbon monolayer. But there were relatively few researchers working in this sphere when the Dresselhauses entered in the mid-1960s. Two of them, Theodore "Ted" Geballe and Bruce Hannay of Bell Labs in New Jersey, approached Millie one day after a talk she gave. In prior work, Geballe, Hannay, and several of their colleagues had made a fascinating discovery: when they inserted alkali metals such as sodium or potassium into graphite layers, the entire system became superconducting— that is, the graphite lost all of its electrical resistance. A bit like breeding a beagle with a Dalmatian and getting a fox, it was a mysterious result, considering that none of the individual components that made up the final compound were super-conducting themselves.[12]

Knowing of Millie's work in magneto-optics, Geballe and Hannay suggested that she look into their head-scratching superconducting compounds. Millie was certainly intrigued; after some of her earliest work investigating graphite as a bulk material, she had become curious to find out what would happen if the individual graphene layers could be separated. The Bell Labs work raised a related fundamental question: What happens to the electronic properties of graphite when you weave individual graphene layers together with different materials?[13]

This line of inquiry would serve as the basis for over fifteen years of research in Millie's lab, including some two dozen theses. She and her colleagues and students investigated every property they could think of by combining graphite with dozens of other atoms and molecules—magnetism, supercon-ductivity, the works. In time, this work would lead to advances in the engineering of new and improved applications, such as the rechargeable lithium ion batteries in your cell phone.[14]

But it would take several years for Millie to begin translating a desire to know into active investigation. As she recalled in a 2001 interview, she was stumped early on by how to design the initial tests: "I didn't have an idea of what the experiment should be, so . . . I put it somewhere in the back of my mind."[15] When a colleague's paper helped to clarify for her how to proceed with examining graphite intercalation, she and a student—Deborah Chung, now a highly accomplished profes-sor of materials science and engineering at the University at Buffalo—finally dove in in 1973. Their first experiments, start-ing with work on graphite intercalated with bromine, were aided by her appointment to the Rockefeller Mauzé chair, which provided a small purse for scholarly research.[16]

Today, isolating graphene is a fairly routine process, although a number of challenges remain on the road to mass-producing high-quality samples. One of the most popular methods for manufacturing graphene flakes is chemical vapor deposition (CVD), a technique in which a base material is exposed to a gas, leading to the formation of a thin layer on top of the base. In the case of graphene, a carbon-containing material is first heated to extreme temperatures so that the material breaks down and carbon atoms come free. This is done on the surface of a metal—commonly copper—which keeps the carbon atoms from flying off as soot. Copper is used as a catalyst to induce the loose carbon atoms to react together and form a layer of graphene on top of the metal. Finally, the graphene must be carefully exfoliated, or removed, from the metal on which it forms.[17]

Back in the 1970s, though, intercalation was the only avail-able method for jimmying graphite layers apart from each other. Using primarily X-ray diffraction, the technique that British crystallographer Rosalind Franklin famously employed to approximate the structure of DNA, the investigation into what happens when guest species are added to graphite turned out to be enormously fruitful for the researchers involved.[18]

Millie and her colleagues discovered, for example, that adjusting the spacing in between graphite layers could produce a multitude of special effects, electronic and otherwise. "By varying the intercalant species and concentration," she wrote in a 1984 *Physics Today* article, "one can prepare a large number of compounds with different properties."[19]

Some graphite intercalation samples exhibit one-to-one layering. For every carbon layer there's an accompanying layer of "guest" material, or intercalant (figure 7.1). Such compounds are known in the field as stage 1 compounds. Samples can also have two, three, four, or more carbon layers in between layers of the guest material. From 1973 to about 1990, Millie's group studied the properties of intercalation compounds up to stage 11—that's eleven carbon layers for every layer of intercalant and a very fat piece of graphite pastry.[20]

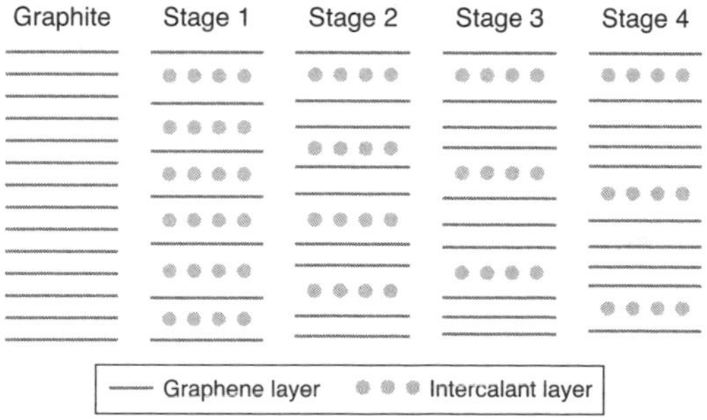

FIGURE 7.1
In graphite intercalation compounds, layers of carbon atoms are interlaced with other elements.

Initially, Millie's team focused on the electronic structure of various graphite intercalation compounds. But they soon became consumed with investigations into graphite host layers—looking at everything from magnetism to electronic properties to superconductivity.[21]

Because it involved two-dimensional materials just a single atom thick, Millie often said that her graphite intercalation work represented her first direct entry into the nanoworld. Individual carbon atoms are only about one-tenth of a nanometer wide. To get a sense of this scale, consider that the average red blood cell is approximately 7 micrometers wide, and a human hair is about 100 micrometers wide—or about 7,000 and 100,000 nanometers, respectively. Science at the nanoscale—dealing with matter roughly 1 to 100 nanometers in size—has exploded in the past couple of decades, but Millie was one of a relatively small number of scientists working with such tiny materials in the 1970s.[22]

"Before 1970 research on graphite intercalation compounds was dominated by chemists and physical chemists," wrote physicist Hiroshi Kamimura in a 1987 review article on graphite intercalation research in Japan. One reason interest picked up among physicists and engineers in the 1970s, Kamimura notes, was the energy crisis facing a number of developed nations. In light of petroleum shortages, citizens of many countries started thinking for the first time about ways to save energy.[23] "We had people from five different academic departments working on this kind of stuff," Millie said in 2013, "and that was another richness of the field."[24]

A "NORMAL" LIFE

As her intercalation work intensified, new opportunities continued to present themselves in other realms of Millie's life. In 1974, she stepped down as associate head of electrical engineering and computer science at MIT when department head Louis Smullin completed his term. (During her tenure, the department added a computer science component that remains today.) Millie's service work would continue elsewhere, however: in that year, she was elected to the National Academy of Engineering (NAE) for her "contributions to the experimental studies of metals and semimetals, and to education." She was only the third woman to be so honored, after human factors engineer Lillian Gilbreth in 1965 and computer scientist Grace Hopper in 1973. As an NAE member, Millie volunteered her time as chair of the Committee on Evaluation Panels at the Bureau of Standards, today known as the National Institute of Standards and Technology.[25]

By the mid-1970s, Millie and her groundbreaking career had also become a subject of public celebration. In June 1976 she was awarded her first of nearly forty honorary degrees, from Worcester Polytechnic Institute, just about an hour's drive west of Cambridge.[26] Two months earlier, an article in *Cosmopolitan* provided the magazine's mostly young, mostly female readership with details about Millie's life and work alongside two other women in engineering (and a very prominent advertisement for Tall Menthol cigarettes). "Have you always thought of engineering as a he-man profession?" the article asked. Millie, for her part, attempted to dispel the myth. "Professional women are normal," she opined, "we want the joys of life just like anybody else."[27]

While stating in *Cosmopolitan* that "I consider my home a career, too,"[28] Millie went above and beyond most others in terms of integrating her children into her laboratory environment—a practice that would inform future activities with her students and also influence physics pedagogy in higher education. When they were preteens, she brought two of her sons into her research sphere at MIT, where she introduced them to her graduate students and had them work odd jobs

PLATE 1

The former Sands Street transit station (top) in Brooklyn, New York, looms over the neighborhood where the Spiewak family lived and worked when Mildred Dresselhaus was a young child. Her parents' former residence, at 45 Sands, was on the north side of this block, at right in the photo. With these structures mostly now gone, this location, at the confluence of Downtown Brooklyn, Brooklyn Heights, and DUMBO, looks entirely different today.

PLATE 2

Baby Millie Spiewak enjoys a stroll with her father, Meyer, in New York City, 1931. Photo courtesy of the Dresselhaus family.

PLATE 3
Millie Spiewak (middle row, center) poses for a photo with her middle school classmates in the Bronx, New York, 1945. Photo courtesy of the Dresselhaus family.

HELEN SCHWARTZMAN
When through *Annals* we do thumb
And to Helen's picture come,
We shall remember that this belle
Whose verses of this kind so well.

DOROTHY SILVERMAN
Novel ideas in every way—
That's the Dorothy of today.
Will she change or stay as now?
That she'll be someone we'll avow.

LUCILLE SIMON
In choosing a gift for Lucy
We must consider these
A palette, a brush, a set of paints,
And a group of beautiful trees.

LEILA SINGH
Leila's vibrant nature,
We know where it lies—
In her smooth, black hair
And in her sparkling eyes.

CATHERINE SMITH
Catherine is a quiet miss,
No sound from her is heard,
Yet fruitful search unearthed a
    voice
As lovely as a bird's.

GLORIA SOLORZANO
Adviser, counselor,
Funster, friend—
In Glory you'll find
The perfect blend.

MILDRED SPIEWAK
Any equation she can solve;
Every problem she can resolve.
Mildred equals brains plus fun,
In math and science she's second to
    none.

PATRICIA STARK
Here's a girl we want to trust,
Of her the Senior Class can boast.
We admire her, we'd like to say,
Because she's sweet in every way.

PLATE 4
Millie's high school yearbook photo and description hint at her exceptional future. Image from the Hunter College High School Annals, Jan. 1948, courtesy of the Hunter College High School Library.

PLATE 5

Gene and Millie Dresselhaus hike on their honeymoon in the summer of 1958. Photo courtesy of the Dresselhaus family.

PLATE 6

Gene and Millie picnic at Cornell University. Photo courtesy of the Dresselhaus family.

PLATE 7

As an infant, Millie and Gene's daughter Marianne joined Millie in the lab while she worked. Photo courtesy of the Dresselhaus family.

PLATE 8

The Dresselhaus family on an outing in Connecticut, 1965. Pictured, left to right: Carl, Paul, Eliot, Millie, and Marianne. Photo courtesy of the Dresselhaus family.

PLATE 9
Millie's parents, Ethel and Meyer Spiewak, visit with the Dresselhaus family. Photo courtesy of the Dresselhaus family.

PLATE 10
Millie and the Dresselhaus children enjoy a day at Yellowstone National Park. Photo courtesy of the Dresselhaus family.

PLATE 11

Millie helps her son, Paul, practice the violin. Photo courtesy of the Dresselhaus family.

PLATE 12

Professor Dresselhaus gives a lecture at MIT. Photo courtesy of the MIT Museum.

PLATE 13

Professor Mildred Dresselhaus at MIT, 1973. Photo by Margo Foote/
MIT News Office, courtesy of the MIT Museum.

PLATE 14
Millie plays in a quintet with MIT electrical engineering graduate students (left to right) Bernd Neumann, Stephen D. Umans, Andrew C. Goldstein, and Alan J. Grodzinsky. Photo courtesy of the MIT Museum.

PLATE 15
Millie (center) and Gene host an informal gathering of students at their home in 1975. Photo courtesy of the Dresselhaus family.

PLATE 16

Millie congratulates her daughter, Marianne, upon graduating from MIT in 1981. Photo courtesy of the Dresselhaus family.

PLATE 17

U.S. President George H.W. Bush honors Millie at the White House National Medals of Science and Technology award ceremony in 1990. Photo courtesy of the George H.W. Bush Presidential Library and Museum.

PLATE 18

Millie meets with her mentor, the Nobel Prize-winning medical physicist Rosalyn Yalow, in 1991. Photo courtesy of the Dresselhaus family.

PLATE 19

With Gene by her side, Millie is sworn in as director of the U.S. Department of Energy Office of Science in 2000. Ernest Moniz, longtime MIT professor and secretary of the Department of Energy under President Barack Obama, is seated at right. Photo by Donna Coveney/MIT.

PLATE 20
Millie plays violin with grandaughter Elizabeth Dresselhaus. Photo courtesy of Elizabeth Dresselhaus.

PLATE 21
U.S. President Barack Obama greets 2012 Enrico Fermi Award winners Millie (with Gene) and Burton Richter (with wife Laurose) at the White House. Official White House photo by Pete Souza.

PLATE 22

Millie stands on stage after receiving the prestigious 2012 Kavli Prize in Nanoscience in Oslo, Norway. Photo by AAS, ERLEND/AFP/Getty Images.

PLATE 23
During her years as Institute Professor Emerita at MIT, Millie remained active as a researcher and mentor. Photo by Bryce Vickmark.

PLATE 24
Millie poses in her office with a minifigure likeness custom-designed by the author. Photo by the author.

PLATE 25

Millie receives the Presidential Medal of Freedom from U.S. President Barack Obama in 2014. Photo by Pablo Martinez Monsivais/AP Images.

PLATE 26

Millie and grandaughter Shoshi Dresselhaus-Cooper visit the grave of Millie's grandmother in 2016. Photo courtesy of the Dresselhaus family.

PLATE 27
Science writer Theresa Machemer created amigurumi of physicists
Shirley Ann Jackson (left) and Millie Dresselhaus for an MIT course
on the history of women in science and engineering taught by the
author in 2017. Photo by Theresa Machemer.

PLATE 28
"Mildred 'Millie' Dresselhaus w/ Fullerenes," 2014, digital photo by
the author.

PLATE 29

"Physicist Mildred 'Millie' Dresselhaus, the Queen of Carbon Science with Nanotube," 2019, linocut print by artist Ele Willoughby.

PLATE 30

"Mildred Dresselhaus," 2017, digital portrait by artist Gillian Dreher.

alongside them, including tasks related to their research. "That experience," she said in a 2007 oral history interview, "helped me to appreciate what [early research opportunities] could do for even pre-college students. . . . It is an organized way to develop tinkering. And it's good for the graduate students to have some younger buddy to supervise and push around a little bit. Good family-type building."[29] Later, Millie started hosting regular check-ins, family-dinner style, with her students—a common practice today but a novelty back in the 1970s.[30]

Overall, as she recounted in a 2002 interview, Millie viewed raising children while pursuing science as "not such a hard thing for me"[31]—though it bears repeating that she regularly credited both her nanny and the strong support of her husband for helping to make it all possible. This included Gene's taking on an equal share of household and child-rearing responsibilities, as well as his delight at the fact that his wife had started to become the more prominent among them at work.[32]

"Millie and Gene worked together in all aspects of their lives," their son, Paul, affirms: "In the house, Gene would help out like few men of his generation. . . . And in much of their work together, Gene would work on the theory portion and Millie would work on the experimental side and bringing the two together. One of Millie's greatest strengths was building teams of talented people, and Gene was always the lieutenant to Millie's captain."[33]

Naturally, such a strong union influenced their children as well at every step of their development. Millie admitted in a 2002 interview that perhaps she and Gene brought their work home a little too often. "When our children were small," she said, "they would use [science] phrases that they had heard from us talking to each other, having no idea what they meant."[34]

Paul, who built his own successful career in physics, confirms, "From an early age, I would hear conversations about

'Fermi surfaces' and 'Fermi energies.' I never really planned to go into the same field as my parents, but it sort of came naturally to me with all of the exposure."[35]

Millie's initiative in particular could be hard to turn off, and sometimes her children could feel pressure as a result. "It's hard to be her son or daughter," says son Eliot, a retired computer scientist. "She was a very strong person—not in a difficult way but just . . . if I feel tired today, I don't feel like doing this or that, that never happened to her. Well, it probably did, but she just had so much drive."[36]

In 1977, Millie took on a new leadership role at MIT, one that would allow her to flex her already considerable administrative muscles, this time in a way that focused squarely on research. As director of the MIT Center for Materials Science and Engineering (CMSE), she was initially charged with saving the center from termination. According to Millie, the National Science Foundation had concerns about the program and threatened to end its funding, which would have effectively canceled its existence. Perhaps influenced by her experience bringing her children to the lab with her and letting them work alongside her students, Millie immediately realized where others hadn't that there was little opportunity for advancement or even training of young people—which led to problems for both up-and-coming researchers and the center itself when older leaders eventually left.[37] "My idea was to build up the center, get it in good healthy form, and develop new talent that could be used for other administrative jobs," she said in a 2007 interview.[38]

Her strategy worked: Millie righted the ship in short order and remained its director for six years. The center continued its interdisciplinary materials work for four decades, until it joined a related entity in 2017 and became the MIT Materials Research Laboratory.[39]

For Millie, directorship of the CMSE coincided with another important life change. In 1976, Gene Dresselhaus decided to leave Lincoln Laboratory to accept a research appointment on the MIT campus. During the decade in which the two worked (mostly) apart—he in Lexington, she in Cambridge—their research collaborations had continued but were limited. Now, a confluence of events conspired to rejoin the Dresselhauses more directly in the professional sphere. One was simply that after sixteen years at Lincoln, Gene was looking for a change of scenery. The more pressing issue was that Millie could really use his help; among other endeavors, the new CMSE directorship promised to be a massive undertaking in addition to her regular research program, and she was feeling a bit of heat regarding how much she'd be taking on. So Gene joined Millie's group, though he was never a direct report.[40] "Somebody else is his official boss," she explained in a 2002 interview. "Our relation is that he can always pick whatever problem he wants to work on."[41]

And so it came to pass that the Dresselhauses turned into a materials science power couple. While they'd certainly collaborated on research before the move, their renewed proximity at MIT enhanced their respective projects, and through the rest of their careers, they worked together a significant amount of the time. Like an All-American pitcher and her most trusted catcher, the pair complemented each other in countless ways: Millie as an experimentalist and Gene as a theoretician; she as a gregarious leader and educator and he as a behind-the-scenes workhorse and cheerleader; she as a prolific writer and he as a steadfast editor and sounding board. "It's . . . a lot more pleasant that somebody close to you understands all your craziness," Millie quipped in a 2002 oral history. "Devotion to science as we do it is a kind of craziness."[42]

Let's take a brief pause to appreciate the unlikely reversal of roles that ultimately took place as the Dresselhauses pursued their parallel careers. Gene had been the more prominent physicist by far when he began as an assistant professor at Cornell and when the couple first moved to Lincoln Laboratory. But by the 1970s, and certainly by the turn of the twenty-first century, it was Millie who had become one of the most famous faces of condensed-matter physics, despite the fact that she and Gene coauthored many papers and books. In the late 1950s and early 1960s, when the Dresselhauses began their careers, many male scientists would have felt humiliated at the idea of his wife outshining him at work. By all accounts, Gene's attitude was as far from that as night is from day. He was always Millie's biggest fan and wanted nothing but the best for her and her career.[43]

"Apparently Gene was very outspoken about his dislike for certain scientists who refused to take Millie seriously, especially in the early days," says granddaughter Elizabeth Dresselhaus. "He was always on her team, working with her and their graduate students every day."[44]

To be sure, Millie earned a significant portion of her renown by way of interactions with thousands of individuals—not only through research endeavors but through her administrative and advocacy work. Yet Millie and Gene collaborated as a research pair for the duration of their forty years together at MIT and, unusual for a scientific couple, Gene has little to show for it in the way of honors or recognition. As of this writing, he hasn't even merited his own Wikipedia article. "For many of the things we did together, people give me more credit than him; I don't think that this is fair," Millie noted in 2002. "I never would be doing what I'm doing now without him."[45]

A DISCIPLINE TAKES ROOT

With a prominent stone castle poised in the center of town and numerous piers studded with gleaming white speedboats, yachts, and other sailing vessels, the French municipality of Mandelieu-La Napoule, located southwest of Cannes on the French Riviera, may seem like an unusual place in which to begin a scientific movement. But in 1977, researchers from around the world who had been experimenting with graphite intercalation compounds decided to convene there for the first time to discuss progress, sticking points, and new directions for their work. It would be the first of many gatherings, hosted in cities around the globe, to serve as a forum on graphite intercalation.[46]

Because most of the attendees of this first conference were new to the field, the event served as an important moment for establishing recent advances within chemistry and physics related to intercalation. Millie and Gene were among those who considered themselves highly enlightened by the meeting, so much so that they used it as a jumping-off point to craft a highly influential 176-page review of what was known in the field to that point—"a small book, like a novelette," Millie described it in a 2002 interview.[47] The meeting also served as a spark for igniting new international research collaborations. This and subsequent conferences led to lifelong associations between Millie's group and research teams around the globe.[48]

In the end, over a decade and a half of work, graphite intercalation compound investigations conducted by Millie, Gene, their students, and colleagues near and far led to a host of important conclusions. They learned, for example, that graphite intercalation generally leads to substantial changes in the concentration of available electrons compared with pristine graphite—and this, in turn, affects the flow of electricity through the affected compound. Intercalation also considerably alters graphite's magnetic properties, with its modified behavior depending on whether the guest material is itself magnetic. And, in response to Ted Geballe and Bruce Hannay's original findings relating to superconductivity, intercalation compounds were confirmed to display unusual superconducting properties.[49]

Among their most significant insights, Millie and her group also determined that stage 1 compounds—single layers of graphene with atoms of an intercalate on either side—are unique in their electricity-conducting properties, unlike anything seen in higher stages. But studies of individual compounds also gave rise to novel findings, like the fact that certain compounds behave like catalysts, which serve to jump-start chemical reactions, while other materials exhibit what Millie called "quasi-two-dimensional" structure and properties— functionally two-dimensional surfaces, considering their infinitesimally small cross sections.[50]

Together, these results foreshadowed much of today's focus on graphene as a promising material for applications requiring high-efficiency electricity conduction—such as fuel cells, which create electricity from hydrogen; lithium ion batteries, which pack an enormous punch in terms of energy density; and supercapacitors, high-capacity energy storage devices that should definitely not be confused with the central operating mechanism of Delorean time machines.[51] "[Millie's] early work on graphite contained most of what is now rediscovered in the case of graphene," Phaedon Avouris, a condensed-matter physicist at IBM, noted in a 2013 *MIT Technology Review* article.[52]

While Millie remained focused on intercalation compounds until around 1990, other subjects caught her interest in the intervening years—and they eventually became the work for which she is best known. When she gave talks toward the end of her career, reviewing her life's work on carbon and other semimetals, she often began with a chart detailing the relative popularity of carbon-related terms in published papers over her fifty-plus years in the business. Graphite intercalation was an early subfield that gained momentum slowly and steadily, but as we are about to see, a number of carbon allotropes would soon capture the imaginations of not only Millie but of a vast international network of physicists, chemists, and materials scientists and engineers. This frenzy would lead to a dramatic uptick in activity, including hundreds of research articles and patents related to carbon in its newly discovered forms.[53]

8   CARBON ZOO

Flexible, paper-thin, and wearable electronic devices that curve with any surface. Energy storage systems that drastically improve efficiencies of devices like batteries and supercapacitors. Improved filtration membranes that help provide cheap, easy access to potable water from brine. Next-generation composite materials that make products from planes to sporting equipment lighter and more resistant to fracture. Futuristic body armor that brings superhero suits—able to resist speeding bullets—into reality.

In the 1980s and 1990s, a zoo of new carbon structures brought to the fore a seemingly endless wish list of inventions that promised to bring materials of science fiction into our everyday lives. Some of these ideas have borne fruit and are already out in the world. Others remain inspirations for many of the scientists and engineers working to uncover new secrets of carbon and to find ways of weaving them into the next generation of wonder materials.

For her part, Millie, the insatiable scientific explorer, continued her research into graphite intercalation compounds for as long as her students remained interested. But starting in the 1980s, novel, seemingly magical forms of carbon caught her imagination and would ultimately pull her, Gene, and her students away on new research adventures.[1]

One such material was carbon fiber, a superstrong, light-weight class of filaments that for many has become an inescapable part of daily life. When we fly, we benefit from the durability of carbon fibers, which have been used to make lightweight parts in the aerospace industry for decades. If you're a tennis player, golfer, cyclist, skier, fisher, or biker, your gear may also be reinforced with a type of plastic known as carbon fiber reinforced polymer, or simply carbon composite.

Carbon fiber can also be found in certain structural materials. The modulus—a measure of an object's resistance to stress—for typical commercial carbon fiber in lightweight structures is slightly higher than the modulus of steel. But a measure known as specific modulus (the modulus divided by the density of the material) is much higher for carbon fiber than for steel, making carbon fibers very attractive for lightweight structures. Carbon fibers are too expensive to use for reinforcing large structures that don't need to be lightweight. But thanks to their electrical conductivity, short carbon fibers can be incorporated into concrete so that it senses strain and damage. This is known as "smart concrete," and it was invented in 1993 by Deborah Chung, the first woman to earn a PhD under Millie's guidance at MIT.[2]

Carbon fibers were first described in the late 1800s, when a handful of inventors used them as filaments for early incandescent light bulbs. As a child, you may have learned that American inventor Thomas Edison created the first incandescent bulb; in reality, British inventor Joseph Swan probably has a more legitimate claim to that feat, while Edison's contribution was in developing the first practical incandescent bulb. Key to this accomplishment was carbonizing fibers of cotton and Japanese bamboo, which were much more reliable than the fatter carbon rods Swan had been using. One-upping both Swan and Edison on carbon fiber, however, was the impressive Lewis Latimer, one of the few prominent African American inventors of the nineteenth century. In 1881, while working for the United States Electric Lighting Company, Latimer developed and patented an improved carbon fiber that lasted significantly longer than Edison's, leading directly to affordable lighting for the general public. Though Latimer never gained the recognition that his former rival (and later employer) Edison did, he was inducted into the National Inventors Hall of Fame in 2006 for his "Durable Carbon Filament for Electric Light Bulbs" (US Patent No. 252,386) and other works.[3]

The transition metal tungsten would supplant carbon as the predominant filament material in incandescent light bulbs before long. But In the 1950s and 1960s, carbon fibers were once again a hot subject when labs around the United States became highly invested in basic materials research for potential applications in aerospace and other industries. This new tide began more or less in 1958 when Roger Bacon, a physicist at the Union Carbide Corporation, developed high-performance carbon fibers in the form of "whiskers"— graphene sheets rolled up into tiny scrolls, which he described in a highly influential paper two years later.[4]

Today, carbon fibers take various forms, depending on how they're created and on the precursor material from which they are synthesized. So-called conventional carbon fibers are the most commonly used for commercial purposes, and they're produced by squeezing an organic resin through a microscopically small nozzle—like the tiniest tube of toothpaste you can imagine. Precursors including rayon, a polymer made from cellulose fiber, and polyacrylonitrile, a synthetic polymer resin better known as PAN, become "carbonized" when exposed to high heat, generally 1,000 to 2,500 degrees Celsius. In this process, the precursor atoms react to become mostly or all carbon, sometimes in a highly organized graphite-like structure of layered sheets, other times in a haphazard mishmash of overlapping sheets. The result is a fiber with a diameter ranging from about 7 to 20 micrometers (7,000 to 20,000 nanometers).[5]

Vapor-grown carbon fibers, which range in size from about 0.01 to 15 micrometers (10 to 15,000 nanometers) in diameter, are created through the same general process described in chapter 7—chemical vapor deposition, in which graphene is grown by exposing a substrate to a gas of free-floating carbon atoms. If you squint really hard, you might be able to see a conventional carbon fiber or a vapor-grown carbon fiber without a microscope, but they look a lot more impressive when viewed with the help of imaging technology (figure 8.1).[6]

Carbon nanofibers are approximately 10 to 150 nanometers (0.01 to 0.15 micrometer) in diameter. Some carbon nanofibers resemble stacks of tiny coffee filters, while others look like minuscule Panda licorice: rods with multiple hollow pockets running through the middle. Like conventional carbon

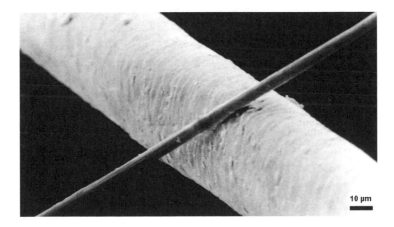

10 µm

FIGURE 8.1

A 6-micrometer-wide carbon filament is superimposed on a 50-micrometer-wide human hair. Credit: Saperaud~commonswiki /Wikimedia Commons.

fibers, carbon nanofibers are strong and flexible. They have applications in energy storage, materials reinforcement, and even bone tissue scaffolding, among other areas.[7]

Millie and her group entered the carbon fiber world in May 1980 when she met a thirty-five-year-old researcher, Morinobu Endo, who had been studying carbon fibers in Japan. At a conference on Cape Cod, Massachusetts, Millie's mind raced with ideas as she listened to Endo describe his latest results on vapor-grown carbon fibers, which include what we now know as carbon nanofibers and carbon nanotubes. After the talk, she introduced herself and suggested that the two team up to investigate intercalation within carbon fibers and compare results with her previous work on intercalated graphite.[8] "I knew she was already a world-renowned expert on carbon, and I had dreamed of working with her," Endo said of the exchange in a 2017 *Physics Today* tribute. Their professional collaboration (and personal friendship) would flourish for the rest of Millie's life—over thirty-six years and some 150 joint publications.[9]

With Endo as a new collaborator, Millie and Gene dove into carbon fibers and nanofibers and would become experts in the preparation and properties of these materials—which in the 1980s were rapidly being exploited by various industries, including aeronautics engineering firms, for both military and civilian uses. These properties included the fibers' complex structure and the dynamics within their atomic lattices, their astounding strength in comparison to their weight, their thermal and mechanical properties, and their electronic structure (figure 8.2).[10]

As with their lengthy review article on intercalation physics, Millie and Gene eventually produced a review-style book, published in 1988, on carbon fibers and filaments, which benefited students, research scientists, and engineers from wide-ranging disciplines—anyone, that is, needing a primer on the

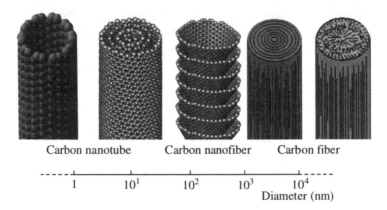

Carbon nanotube   Carbon nanofiber   Carbon fiber

1   10^1 10^2 10^3 10^4
                                                          Diameter (nm)

FIGURE 8.2

With their colleague Morinobu Endo, Millie and Gene Dresselhaus
became experts in carbon fibers, nanofibers, and nanotubes.
Credit: Reprinted from Robert Vajtai, ed., *Springer Handbook of
Nanomaterials* (Springer, 2013).

field. But fibers were just the tip of the scientific iceberg when
it came to new varieties of carbon—and the Dresselhauses
simultaneously began dipping their toes into additional lines
of inquiry relating to forms even more fantastic: buckyballs,
nanotubes, and more.[11]

It was, in fact, through Endo that Millie began to think
deeply about carbon nanotubes, an exciting form of carbon
resembling an infinitesimally small chicken-wire straw, whose
remarkable properties would remain "interesting"—Millie's
favorite phrase to describe topics she considered worthy of
scientific inquiry—for the rest of her life. Shortly after they
met in 1980, Endo relayed to Millie a line of questioning by
theoretical physicist Ryogo Kubo to those gathered at a small
conference of mostly Japanese researchers. "Would it be possible to make a carbon fiber that was just one layer thick?" Millie recalled of Kubo's thought experiment in a 2013 interview

with the Materials Research Society. "That was the concept of the single-walled nanotubes. And this idea sort of settled in my mind."[12]

Nanotubes didn't become an immediate focus, however. Before that, as she continued her work on intercalation and on carbon fibers and nanofibers, Millie began another set of experiments, shooting carbon samples with a laser—a process known as ablation—and recording what flew off. What she discovered as a result of these tests would lead to a whole new field of carbon science—and a Nobel Prize for a trio of chemists.[13]

A NATIONAL LEADER

The 1980s saw Millie's star rise dramatically, thanks largely to her high-caliber research and teaching in the classroom and through technical writings that served to educate and inspire students and colleagues around the world. She had established herself as an exceptional professor at MIT for both her academic output as well as her teaching and mentorship, but in this decade, she would ascend to the highest echelon of the Institute's faculty. During the 1980s, she would also get an opportunity to develop her science leadership skills at the national level—an activity that would become a frequent calling over the following decades.

In 1982, Millie was appointed to two important posts related to major US physics organizations. One was the governing board of the American Institute of Physics (AIP), a promotional and publishing society whose members—today numbering 120,000 scientists, engineers, educators, and students—belong to a variety of physics-related societies around the United States. Millie was one of very few women on the AIP board during this period, and she served for three years, through 1985.[14]

Millie was also nominated for and then elected to the presidential line of the American Physical Society, an AIP member society that at the time represented 35,000 physicists (it has grown to 55,000 today). She held a three-year leadership term with increasing responsibility each year, culminating in the highly distinguished APS presidency in 1984—only the second woman to be tapped for this prestigious role (the first was Columbia University's pioneering nuclear physicist Chien-Shiung Wu).[15] "I could hardly believe that I was a serious candidate," Millie wrote in a *Physics Today* article summarizing her APS experience. "My boss at MIT assured me that the most valuable contribution that MIT could make to women in science was for me to take this proposition seriously."[16]

It was a life-changing experience. In addition to the mind-expanding opportunities to collaborate with numerous individuals across physics disciplines, Millie also played a significant role in helping to shape the community and its overall priorities. She used her term to advance what she called an activist agenda within the society, projects that mirrored some of the greatest hits of her career to that point: support for basic science on issues of national concern; for greater international collaboration among physicists; for the development and advancement of young, up-and-coming physicists; and for projects that ensured improved physics and science literacy among the general public.[17]

Millie also led in supporting roles even after her APS presidency ended, including as chair of the Committee on the Status of Women in Physics. In that position, she was instrumental in establishing the Climate for Women Site Visit Program, an ongoing project, now run in conjunction with the APS Committee on Minorities in Physics, aiming to ensure that college physics departments and independent labs around the United States are keeping up with best practices in terms of supporting underrepresented students and faculty.[18]

According to University of North Carolina professor Laurie McNeil, one of Millie's postdocs in the early 1980s who has also been active within the APS, these site visits "weren't just vital to changing the climate at many universities; the program, itself, was a wonderful learning experience. . . . It was highly instructive to listen to Millie deliver difficult news to a department chair or division head, but do so in a way that was likely to create institutional change, instead of just creating additional resentment."[19] As an example of such tactful message delivery, McNeil noted that Ball State University professor emerita Ruth Howes recalled one site visit in particular, when Millie "gave the department hell over its track record with women. Politely, of course."[20]

Back at MIT, things were humming for Millie and Gene. In 1983, in what was likely a response to the awkward optics of Millie being selected to lead one of the nation's preeminent physics organizations without belonging to her own institution's physics department, she was awarded a joint appointment in the MIT Department of Physics, where she remained (in addition to her original home of electrical engineering) for the rest of her career.[21]

Overall, the 1980s continued to be a busy time for the Dresselhauses. By then their children were all in their late teens and early twenties. Marianne, the eldest, was at MIT as an undergraduate, double-majoring in mechanical and nuclear engineering. "It wasn't that my parents were there," she says. "I wanted to go . . . MIT's an exciting place!"[22] Paul also became a physics and electrical engineering student at MIT, while Eliot studied math at Harvard.[23]

GREAT BALLS OF CARBON

On the first of October in 1982, a brilliant seventeen-story geometric sphere stood out from a stand of palm trees, poised to become one of the most recognizable structures in the world. It was the opening day of Epcot Center, a theme park in Bay Lake, Florida, based loosely on entrepreneur Walt Disney's vision for a "community of tomorrow" that would serve as a test bed for new ideas in city planning. At the center of the park stood Spaceship Earth, gleaming with 11,324 facets, in front of which singers, dancers, and instrumentalists clad in white 1980s-era "space explorer" outfits performed as representatives of the Walt Disney Company and the state of Florida welcomed the first visitors and, by way of television, the entire world to its newest offering. Inside, the sphere housed a family-friendly ride that treated guests to a fifteen-minute history of communications, from paintings of cave dwellers to a quotidian home computer of 1982.[24]

Both Epcot and Spaceship Earth as its centerpiece became family favorites. In recent years, the park has hosted around 12 million visitors annually, though attendance dropped precipitously in response to the COVID-19 pandemic. Yet many today are unaware of the man who served as the inspiration for the great ball at the center of it all: American inventor and architect Richard Buckminster Fuller.[25]

Born in the last decade of the nineteenth century, Fuller was a two-time Harvard flunkee whose visionary thinking proved too off-the-wall for some. But Fuller didn't really care what anyone thought of him; he was too busy reinventing architecture with the insights of a systems engineer and the design sense of a mathematician. He popularized concepts, many of which he invented, that sound like something out of an Aldous Huxley novel (in fact, some influence may have gone both ways, as the two were acquaintances): *synergetics, tensegrity, dymaxion, livingry*, and *Spaceship Earth*, among others. Fuller used the latter in his "Operating Manual for Spaceship Earth," a 1968 article in which he equated our home planet with a spaceship of limited resources and prognosticated what could happen if humans ceased to take care of it.[26]

But Fuller is perhaps best known for his championing of the geodesic dome, an architectural structure based on geometric solids known as geodesic polyhedra, which feature repeating triangular sides, or facets. When Disney "Imagineers" were dreaming up the structure that would house their new theme park's signature attraction, it made perfect sense that they would look to a futurist like Fuller and his decidedly cosmic designs—and that for its name, they'd borrow a term he (and others) had used to invoke a new age that promised great technological advances but also threatened to diminish precious resources.[27]

Though they never fully caught on as the edifice of the future, spherical geometric structures became synonymous with Buckminster Fuller ("Bucky," as he was widely known). So when a new form of carbon in the shape of a truncated icosahedron—a geometric soccer-ball-like sphere made with sixty unique vertices and repeating pentagon and hexagon faces—was discovered just a few years after Epcot opened its doors, scientists quickly recognized an opportunity to pay homage to a great thinker: with the naming of the buckminsterfullerene, or buckyball, that would jump-start decades of research into new possibilities with element number 6 (figure 8.3).[28]

As is true of countless advances in science, the discovery of buckyballs—and the larger class of hollow, spheroid molecules known as fullerenes—was the culmination of work from numerous individuals around the world over many years. In his Nobel Prize speech commemorating the buckyball's discovery, chemist Richard "Rick" Smalley of Rice University noted that the fullerenes came about "as a result of decades of research and development of methods to study first atoms, then polyatomic molecules, and ultimately nanometer-scale aggregates."[29]

(a)   (b)

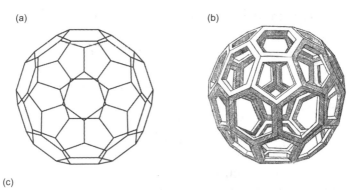

(c)

FIGURE 8.3
Beautiful balls. (a) Buckminsterfullerene, C60. (b) Archimedes
truncated icosahedron by Leonardo da Vinci, published in 1509. (c)
Spaceship Earth at Disney's Epcot, inspired by Buckminster Fuller's
geodesic domes. Credit: For Spaceship Earth, Chensiyuan/Wikimedia
Commons.

In some ways, these foundations go back millennia. The geometry of the truncated icosahedron has long been connected to Archimedes, the great mathematician who lived during the third century BCE. It was also well known to scholars including Piero della Francesca and Leonardo da Vinci in the early 1500s, but the idea of molecules in this configuration wasn't explored until centuries later. According to a 1996 history written by Millie, Gene, and Peter Eklund, Hungarian scientist László Tisza (who would later join the MIT faculty) was one of the first to contemplate icosahedral molecules in the early 1930s. In the 1970s, Japanese and Russian researchers began to consider the possibility of such a molecule composed specifically of carbon, but their work went underappreciated outside their local scientific communities.[30]

Things picked up considerably in the 1980s when several groups began investigating mysterious atomic clusters that were found to coalesce in an inert gas when carbon samples were irradiated by laser light. Of the researchers looking at these clusters, many became focused on characterizing those made of just a handful of atoms—but Millie and her collaborators deduced correctly that the majority of clusters they were seeing in these laser studies had to be relatively large. She suspected they contained many tens of carbon atoms each, possibly upward of one hundred atoms.[31] "Many people laughed at that," Millie said in a 2002 interview. "They thought it was impossible."[32]

What were they seeing that others weren't? For one thing, while their lasers transmitted a relatively modest amount of energy, quite a bit of graphite was being used up during the course of the experiments. As a result, Millie and her collaborators figured that large chunks, rather than smaller bits, must be flying off. There was a more readily apparent reason for their thinking: during their experiments, the researchers' clothes became heavily covered in soot, a blackish residue of burning organic compounds—and a form of carbon that we now know can contain fullerenes.[33]

Millie was soon invited by the Exxon Research and Engineering Company, a research arm of the multinational oil and gas firm, to give a talk about her work on carbon clusters. At the time, the company's scientists were among the world's experts on such clusters as they sought potential applications within the energy industry. They'd previously been working with relatively small carbon clusters, perhaps ten to fifteen atoms large, but when Millie came to visit, she implored them to think bigger—to look into clusters three, four, and five times that size because only larger clusters would explain her results.[34]

They took her advice. In a subsequent experiment using mass spectrometry, a technique that measures the mass of a system based on charged particles known as ions, the Exxon researchers produced a highly influential spectrum that looked at carbon clusters containing upward of one hundred atoms. Published in 1984, their data resembled the undulating vertebrae of an ancient *Spinosaurus*, with a major spike at sixty atoms and a smaller one at seventy atoms. We now know these spikes represent C_{60}, buckminsterfullerene, and the related C_{70} fullerene (figure 8.4).[35] "Whenever there is a big discovery, there are a lot of side discoveries that lead to it," Millie noted in 2002.[36]

To be sure, although signals from these mystery molecules were already staring back at them, the fullerene spikes weren't pronounced enough for the researchers at Exxon—or elsewhere, for that matter—to truly see them for what they were. Rather, they were a gray ship in a sea of indigo, while scientists actively sought a vessel of neon yellow.[37]

Still, experts in the field could sense change was afoot. At MIT in the mid-1980s, Millie's research attention had become split among graphite intercalation, carbon fibers, and ion

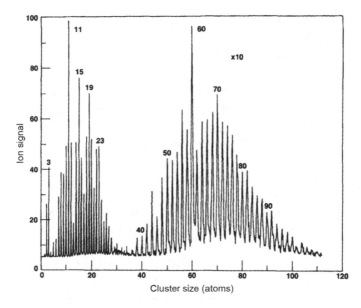

FIGURE 8.4

Scientists at Exxon published this highly influential spectrum shortly after Millie Dresselhaus implored them to search for large carbon clusters. Years later, the spike at sixty atoms was confirmed as the signal for C_{60} (buckminsterfullerene) and the spike at seventy atoms as the signal for C_{70}. Credit: Reprinted from Eric A. Rohlfing, D. M. Cox, and A. Kaldor, "Production and Characterization of Supersonic Carbon Cluster Beams," *Journal of Chemical Physics* 81, 3322 (1984). https://doi.org/10.1063/1.447994.

implantation—the introduction of charged ions into solid materials as a way to affect the materials' properties. After she and her colleagues had discovered the large clusters that resulted from laser ablation of graphite rods, it was others who went on to eventually identify their composition. But spherical molecules weren't completely out of mind during this time. In a story she retold many times, Millie assigned a special problem to one of her classes the year before buckyballs were officially discovered: What would the atomic vibrations look like for an icosahedral molecule made of sixty hydrogen atoms? Her students, the story goes, grew frustrated with Millie, both for not submitting the results for publication and for not assigning them the problem in carbon.[38] "The discovery of fullerenes and nanotubes was in the air," she later told the Kavli Institute.[39]

Meanwhile at Rice University, Rick Smalley and his colleagues were busy with their own independent laser experiments. Focusing at the time on atomic clusters of semiconducting materials, they initially had no desire to compete with the Exxon group's carbon work. But when British physicist Harold Kroto, who had been looking into molecular components of the stellar universe, requested time with the specialized equipment they'd built and were using on the Rice campus, things got interesting very quickly.[40]

Kroto had been searching for a signature indicating the existence of a particular carbon chain that he hypothesized might form inside stars. Early experiments with the Rice team—faculty members Smalley and Robert Curl, plus graduate students James Heath, Yuan Liu, and Sean O'Brien—confirmed the existence of these chains in September 1985. But the experiments were interrupted by what Kroto would later call an "uninvited guest": a chemically stable molecule made of sixty carbon atoms that showed up insistently among their data, like a stray cat that keeps meowing at your door until you finally, one day, let her in. In a most beautiful peak, data they collected using a highly focused experimental setup pointed epically to C_{60} as the dominant source of the carbon clusters that so many others had observed. This work would ultimately serve as a watershed in the discovery of fullerenes, an entirely new family of carbon allotropes whose properties and applications would soon have chemists, physicists, and materials scientists as excited as kids on Halloween.[41]

The structure of C_{60} was not at all obvious at first, but the Rice team proposed an icosahedron with alternating pentagonal and hexagonal rings and named it in honor of Bucky Fuller and his geodesic domes. Five years later, the existence of buckyballs was confirmed when Donald Huffman and colleagues at the University of Arizona and the Max Planck Institute for Nuclear Physics in Germany came up with a way to produce enough of them to investigate thoroughly. Six years hence, and eleven years after their initial work, Curl, Kroto, and Smalley were awarded the 1996 Nobel Prize in Chemistry for their discovery of the fullerenes, which are now considered to include both closed carbon molecules like buckyballs and open-ended carbon nanotubes.[42]

"Fullerenes were one of the first big bolts of energy that helped popularize the field," Ado Jorio, a professor of physics at the Federal University of Minas Gerais and a former postdoc with the Dresselhaus group, noted in 2017, in reference to the now booming area of carbon nanostructures. "But it's important to remember that the fullerene discovery could never have been possible if Millie had not done all the research before it."[43]

Despite the buckyball's early promise, the molecule has been something of an underachiever. C_{60} has found limited roles in biomedicine, but many other promising applications have not materialized. Nevertheless, the molecule's revelation was and is still regarded as a major breakthrough.[44] In part, its discovery helped to usher in the beginning of the nano age: it provided tantalizing support for physicist Richard Feynman's famous 1959 prediction that there is "plenty of room at the bottom"—meaning that infinitesimally small specks of matter, even individual atoms, can be manipulated and employed to create tools with desirable built-in properties.[45] C_{60} also paved the way for the discovery of carbon nanotubes, which to date have had a much stronger impact on the world—and,

as we will see, were a highly fruitful muse for Millie in the latter third of her life.[46]

MILLIE'S MAGIC

Not everyone in science can boast of a fast-paced, decades-
long, fulfilling career. For some, even after early success, their
work may be prone to dead-ends, requiring a constant—and
sometimes exhausting—ability to reinvent themselves. Others, including many who flourish over a decade or two, eventually see their output and impact begin to dwindle as they
age. Then there are the magicians, whose dedication and love
of research translate into one new project after another for
thirty, forty, and fifty years. Endeavors and accolades trickle
in early, until eventually they reach a critical mass that feeds a
never-ending loop of investigation and recognition.

Millie was one of those magicians, and her scholarly point
of no return was undoubtedly the decade between 1980 and
1990. After her service as APS president in 1984 and with the
publication of paper after paper looking into the properties of
carbon and other semimetals, a drizzle of awards and honors
began turning into a steady rain, marking Millie as one of the
most acclaimed scientists in her field—and, eventually, one of
the most distinguished in any field.

Among other accolades, in 1985, at age fifty-four, she was
elected to the US National Academy of Sciences, one of the
highest honors of the scientific profession. At the end of that
year and the following spring, MIT solidified her place as one
of the Institute's most prized faculty members, first by naming
her an Institute Professor, the highest honor bestowed by the
faculty and administration, and then by selecting her for the
James R. Killian, Jr. Faculty Achievement Award for outstanding professional accomplishment.[47] The Killian Committee, in
its recommendation of Millie to the rest of the MIT faculty,

noted in reference to the lecture she would give as part of the award, "We look forward to hearing from Millie how she does it, in hopes that some might rub off on the rest of us!"[48]

Indeed, through the rest of the decade, and into the 1990s, Millie was a scientific workhorse. With her students and colleagues, she published some 175 articles from 1985 to 1990 on everything from ion implantation in polymers and semimetals to superconducting graphite intercalation compounds. She sat on doctoral committees and supported numerous students and postdocs. After years of preparation, she, Gene, and colleagues Ko Sugihara, Ian Spain, and Harris Goldberg published their influential book on carbon fibers and filaments, *Graphite Fibers and Filaments*, in 1988.[49] Millie logged incredible airline miles as she traveled continually for meetings, talks, and conferences. "Every day, every hour, bolts of work would go out," says longtime assistant Laura Doughty. "She was constantly juggling."[50]

She also continued her invaluable work highlighting issues of concern related to women and other marginalized groups at MIT and supporting those communities in their march toward equal access and treatment. As historian Margaret Rossiter noted, Millie in the 1980s worked alongside graduate student Denice Denton to report on gains in terms of more female graduate students entering their Department of Electrical Engineering and Computer Science and on a striking imbalance in the percentage of graduate students who at the time did not complete their degrees—27.5 percent of female students compared to 11.9 percent of male students. She went to bat, too, for female postdocs, who organized for the first time as the MIT Association of Postdoctoral Women in 1988. As Rossiter explains, Millie used funding from her Rockefeller Mauzé chair to support monthly lunches for the group, whose meetings led to the development of an important guide for female MIT postdocs as well as increased negotiating power with the administration on concerns such as health benefits and procedures in the event of harassment.[51]

Outside of her MIT work during this period, Millie engaged in numerous service projects for scientific and engineering societies—for example, as a council member of the National Academy of Engineering and as chair of the Engineering Section of the National Academy of Sciences. She also chaired the National Academies' Committee on Women in Science and Engineering, which authored an important report on increasing numbers of women in STEM in the 1990s.[52] "If you look at Millie's resume, it goes on for seven to eight pages about all the committees that she's served on," Michelle Buchanan, deputy for science and technology at Oak Ridge National Laboratory, noted at a 2017 tribute event. "Millie was one of those rare individuals who was thoughtful and selfless. She was able to step outside of her personal research areas and serve science in a bigger [way]. That's not always a characteristic of high-powered research scientists."[53]

A YEAR TO REMEMBER

As we all do, Millie encountered occasional chasms on her life's path. She was dealt a particularly harsh blow in August 1990 when the National Science Foundation decided to pull funding from MIT's Francis Bitter National Magnet Laboratory and move it to a new facility in development at Florida State University.[54]

In a 2013 *MIT Technology Review* interview, she likened the announcement to what happens when "the bottom falls out of the ship you're on."[55] Millie, her students, postdocs, and others in her group had grown highly dependent on the MIT Magnet Lab, as much of their carbon and other materials research required the use of high-magnetic-field equipment. But there was really no question of moving. While the planned National High Magnetic Field Laboratory would be a monumental, state-of-the-art facility, Millie had no desire to shift her center of operations to Florida.[56]

With support about to evaporate, she was forced to make a major course correction. Looking back on her career, this particular Robert Frost moment—two roads diverging in a wood—was just the latest instance in which her chosen route, for one reason or another, became impeded. Whether it was her thesis adviser refusing to work with her, her supervisor requiring that she abandon her field of specialization, or a strict change regarding business hours at a time when she needed some flexibility, Millie had faced her share of exacting turns. But it was not her style to brood, and the occasion gave her yet another opportunity to put into practice lessons she had learned from her mentor, Enrico Fermi: "You keep a little bit of what you know as your safety position," she told *MIT Technology Review*. "But then you put 90 percent of your effort into starting the new thing that you don't know."[57]

To be sure, Millie wasn't about to give up her carbon focus; indeed, some of her most impactful work investigating carbon and other semimetals was still to come. But she needed to pivot in terms of the ways in which she studied these elements and their interactions experimentally. With her high-magnetic-field days coming to a close, she refocused her studies on carbon's electrons and phonons—particles of heat associated with the vibrational energy of individual atoms—using other methods including spectroscopy, which works by characterizing interactions between matter and light. She also began to consider entirely new lines of research that proved to be quite fruitful.[58]

Things brightened considerably for Millie in the later months of 1990. On November 13, two days after her sixtieth birthday, she and Gene got to dress up for a special occasion: Millie and nineteen other scientists, engineers, and mathematicians from around the United States were awarded the National Medal of Science, one of the highest civilian honors in the nation.[59] She was recognized at the White House by President George H. W. Bush "for her studies of the electronic properties of metals and semi-metals, and for her service to the nation in establishing a prominent place for women in physics and engineering."[60]

Ceremonies for the National Medals of Science and Technology are a delight to watch. Most of the winners are relative unknowns; they may be superstars in their fields, but they're not household names. Millie wasn't the first woman to receive a Medal of Science—nearly a dozen others, including her undergraduate mentor, Rosalyn Yalow, came before her—but she did earn the distinction of being the first woman to win for engineering, one of six award categories recognized by the National Science Foundation, which recommends candidates to the president.[61]

The following month, an exchange at a meeting in Washington heralded the flowering of a new line of inquiry that would enthrall Millie for the rest of her career and help to solidify her reputation as one of the most important scientists in all of solid-state physics. It was a Department of Defense workshop, and Millie had been invited to speak on her carbon fiber work. Rice University chemist Rick Smalley was on hand as well to speak on the fullerenes he'd codiscovered. At a certain point, with both Millie and Smalley onstage, an attendee asked about the connection between their two invited talks: Was there a direct relationship between carbon fibers and fullerenes?[62]

It was a question that harkened back to the one Ryogo Kubo had posed a decade earlier when he wondered about the possibility of a carbon fiber with walls just one layer thick. Millie and Smalley, both familiar with each other's work, dug in. If you added a ring of precisely ten carbon atoms to a C_{60} molecule, you'd get C_{70} and a slightly oblong fullerene. Adding a second ring of ten carbons would yield C_{80}, a third ring would yield C_{90}, and so forth until there was a very long tube of carbon chicken wire, capped at the ends with, essentially, the top and bottom of a buckyball. Millie and Smalley hypothesized that such structures might feature attractive properties.[63] Just one atom thick, they then were called, variously, "tubular fullerenes," "graphitic carbon needles," "C_{60}-based tubules," "graphene tubules," or—as Millie originally dubbed them—"bucky fibers" or "buckytubes."[64] Today, scientists everywhere refer to them as carbon nanotubes, and their early promise has led to numerous applications (figure 8.5).

Millie had officially been bitten by the nanotube bug. The following summer, another quirk conference proceeding would catapult her deep into theoretical studies of nanotubes and would yield one of the most consequential scientific papers of her career. It was a meeting on fullerenes in Philadelphia, and when an expected speaker didn't show up, Millie was asked to fill time with an off-the-cuff talk. She chose to discuss the possible existence of carbon nanotubes, speculating about their properties and impressing on those in attendance that these elongated nanostructures had the potential to become highly important. As she spoke, she witnessed many of her colleagues furiously taking notes.[65]

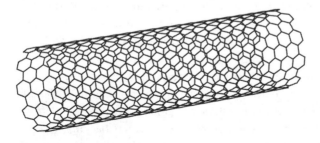

FIGURE 8.5
Carbon nanotubes are essentially rolled-up tubes of graphene.

As Millie predicted, carbon nanotubes' electronic, mechanical, thermal, and other properties led them to being highly favorable for use in a wide variety of applications, making them better, stronger, and faster than existing materials. They are, for instance, highly conductive, both electrically and thermally. They're superelastic and flexible, so they can be stretched and bent without harming the underlying structure. And owing to their powerful chemical bonds, they are incredibly strong.[66]

To continue with the carbon-as-pastry analogies from previous chapters, the closest analogues to nanotubes are probably French pirouettes or Spanish *barquillos*: rolled-up tubes of extremely thin cookie wafer. Carbon nanotubes can be single-walled (SWNTs) or multiwalled (MWNTs), with the latter featuring concentric tubes of graphene rather than one spiraling roll.

Today, the tiny hollow rods might be considered a Swiss Army knife of nanotech because their uses are manifold. A 2018 report found the global carbon nanotube market had already reached $4.55 billion and forecast that the figure would more than double to $9.84 billion by 2023.[67] A sampling of realized and emerging carbon nanotube applications illustrates their import across many fields—for example: mirror housings, fuel lines, and filters in cars and trucks; electromagnetic interference shielding and wafer carriers in microelectronics; epoxy resins; reinforced concrete; sporting equipment; carbon fiber precursor materials; reinforced composites in turbine blades, ship hulls, and aircraft structures; yarns and laminated sheets in bullet-resistant clothing; flame-retardant plastic additives; anticorrosion coatings; transparent electronic films in flexible flat screens and photovoltaics; window and sidewalk defrosters; transistors; OLED displays; LCD screen backlights; RFID tags; microchip memory; a powder widely used in laptop and cell phone batteries for better interconnectivity and longer battery life; fuel cell components; water purification filters; coatings that increase acoustic absorption; gas and toxin sensors; biosensors to help detect DNA, proteins, and hormone levels; cancer therapy; and scaffolding for tissue regeneration.[68]

While many would agree that Millie and Rick Smalley's 1990 conference discussion sparked the discovery of the formation of SWNTs soon after, it bears noting that the overall discovery of carbon nanotubes has a complex history. It may be that a number of groups working with them before the 1990s did not know precisely what they were looking at. Other early work failed for one reason or another to become well known or accepted at the time it was published.[69]

For example, some studying notably larger structures—such as carbon fibers and whiskers—may have, in certain cases, been observing nanoscale MWNTs without realizing it, in part because equipment at the time wasn't able to resolve nanometer-sized tubes. A 1952 paper by Russian researchers L. V. Radushkevich and V. M. Lukyanovich appears to be the first published report of a nanoscale carbon tubule, but it appeared in a relatively obscure Russian journal, in the Russian language, during the heart of the Cold War—three strikes against achieving international attention. And while their work didn't catch on at the time, a set of 1976 papers by Morinobu Endo, fellow Japanese scientist Tsuneo Koyama, and French scientist Agnès Oberlin included what many now consider the first published image of a single- or double-walled carbon nanotube. Millie herself noted that international travel for science conferences was uncommon before 1980—and of course the Internet didn't yet exist then—so the fact that science was happening in bubbles around the world was another contributing factor to some research not gaining traction.[70]

The carbon community is, however, certain about one thing: the late 1991 letter to *Nature* by Japanese physicist

Sumio Iijima reporting unambiguous proof of the existence of multiwalled nanotubes is considered a defining moment in the history of carbon science. Spurred by lively conference proceedings, improved imaging equipment, and a steady stream of new discoveries, carbon nanotube papers began to pop up rapidly after this report.[71]

Another defining moment took place just months later, in 1992, when Millie and a pair of young visiting researchers published key articles on carbon nanotubes that would forever change how scientists and engineers think of these minuscule pipes.[72]

Physicist Riichiro Saito arrived at MIT from Japan in fall 1991, eager to work for an academic year on carbon nanotubes with his host, Gene Dresselhaus. Saito, at age thirty-three, had met Gene and Millie in the 1980s and was excited to collaborate with the couple, who by then were renowned in the carbon field.[73] "Gene-sensei and Millie-sensei always worked as a team and spoke together about everything," Saito said at a 2017 tribute event. "When I [gave] the daily report to Gene-sensei in the afternoon before going home, Millie-sensei would always know everything the next morning and be happy to help."[74]

Nearly simultaneously, another young Japanese physicist who had just arrived at MIT, Mitsutaka Fujita, began attending Millie's group meetings, excited about nanotubes. Millie suggested that the two visitors attempt to solve some nanotube mysteries together. Thus, Fujita, Saito, Millie, and Gene ended up working as a team on two 1992 papers, producing insights that would prove invaluable to the carbon community.[75]

To understand their contribution, let's examine the detailed structure of single-walled carbon nanotubes to see how they roll. SWNTs come in three varieties, each with slightly different atomic geometries (figure 8.6). In the so-called armchair configuration, a nanotube's six-carbon hexagons are oriented

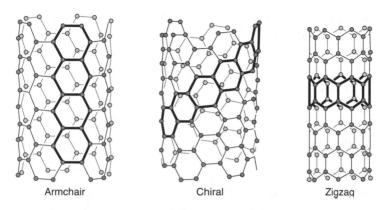

Armchair   Chiral   Zigzag

FIGURE 8.6

Millie Dresselhaus and her colleagues proposed, correctly, that carbon nanotubes would behave differently depending on the configuration of their atoms.

with "tops" and "bottoms" of repeating rings parallel to each other and stacked up and down the long axis of the nanotube. Following the path that an energy current might take from one hexagon to the next, you can make out (squinting hard enough, anyway) what might appear like a chair. A second nanotube structure, zigzag, also features hexagons with perpendicular sides, but instead of stacking up along the long nanotube axis, they hug the nanotube like a ring. The third form, chiral, features hexagons that spin around the tube at a slight angle.[76]

In what are considered the first theoretical predictions of single-walled carbon nanotubes' electronic properties, Millie, Gene, Saito, and Fujita proposed that nanotubes would behave quite differently depending on their diameter and the orientation of their carbon hexagons—that is, whether the nanotube in question had an armchair, zigzag, or chiral configuration. Armchair nanotubes, they posited, should mirror metals in terms of current easily flowing through, while zig-zag or chiral nanotubes should be semiconducting, allowing electricity to pass only when extra energy is added to the system.[77] This work "immediately ignited a world interest in carbon nanotubes," Israeli chemist Reshef Tenne said at a 2018 tribute event.[78]

But this idea—that such wildly different properties would result from a slight difference in a nanotube's atomic orientation—was not something that other physicists readily accepted. In fact, it was fairly controversial, and it would take a number of years before the MIT team was vindicated. "Research communities are always very conservative," Millie stated in 2007. "They like to believe what has happened before is correct. And when you come up with something that's revolutionary and new, they don't like this. . . . Of course, when you have the smoking gun, it's all over."[79]

The smoking gun in this case materialized in the mid-1990s after single-wall carbon nanotubes were successfully observed in 1993 by independent teams at NEC Laboratories in New Jersey and IBM Research in California. The NEC team was led by Sumio Iijima, who had jump-started the field with his article on multiwalled nanotubes two years earlier, while the IBM Research team was led by physicist Donald Bethune, who had previously helped to confirm the structure of C_{60} and C_{70} fullerenes. Once SWNTs were officially discovered, scientists around the world began testing physicists' theoretical predictions of how they would behave—and confirming, in the case of Millie, Gene, Saito, and Fujita, that their earlier calculations had been correct. Of course, Millie and Gene worked diligently to follow up on their own predictions as well. In 1997, they reported their isolation of nearly pure SWNT samples using specialized spectroscopy.[80]

Discoveries relating to carbon nanotubes quickly turned into applications across numerous fields. Today, nanotubes are

FIGURE 8.7
Carbon nanotubes are grown in a "forest." Credit: Reprinted from
J. Li, C. Papadopoulos, and J. M. Xu, "Highly-Ordered Carbon
Nanotube Arrays for Electronics Applications," *Applied Physics
Letters* 75, no. 367 (1999); https://doi.org/10.1063/1.124377.

FIGURE 8.8
Nanotubes are spun into an ultrastrong "yarn." Credit:
Commonwealth Scientific and Industrial Research Organization.

all around us, but scientists and engineers continue to investigate their properties and to improve their production in forms that can be even more useful for processes and products.

In the 1990s, the Dresselhauses would once again take to enlightening others with their writing, this time through an influential book on fullerenes and carbon nanotubes written with physicist Peter Eklund, a longtime collaborator and former postdoc in the Dresselhaus group. But their fascination with these materials didn't end when their writing was done. For Millie and Gene, carbon nanotubes remained a subject of great interest and investigation for the balance of their careers.[81] As Millie stated in 2014, "Now, we are actually able to do many things about nanotubes that I never dreamed would be possible."[82]

9   LEADING BY EXAMPLE

Marcie Black was in a bind. The MIT graduate in electrical engineering and computer science had just returned to the Institute to pursue a PhD, eyeing a career in which she could help to devise new solutions to the energy crisis—the world's rapidly increasing appetite for energy at a time when scientists had become convinced that the consumption of fossil fuels was causing significant environmental harm. It was 1997, the same year in which the world's developed nations gathered to adopt the Kyoto Protocol, a groundbreaking international treaty establishing binding limits on greenhouse gas emissions for individual countries.[1]

Black had been lucky enough seven years earlier when Millie was assigned to be her undergraduate adviser.[2] "I had no idea how famous she was," Black later wrote. "To me, she was someone that knew MIT, cared about my success at MIT, and gave me advice on classes."[3] Now, with a clearer picture of where she wanted to direct her professional energies, Black dropped in to visit her former adviser, whose work fascinated her.

At the time, scientists and engineers around the globe were actively investigating how hydrogen might one day supplant fossil fuels as a clean energy source, especially in the transportation sector. One of the biggest challenges in this area had been determining how to safely, efficiently, and inexpensively store hydrogen as a fuel, so when carbon nanotubes arrived in the 1990s, Millie jumped on the opportunity to consider whether these structures might prove useful in the realm of energy storage. Black was inspired by Millie's work—which had begun to show nanotubes as promising sponges for hydrogen atoms—and hoped to join her group's research efforts for her doctoral work.[4] As Black, now a successful CEO and entrepreneur, remembered in an interview years later, it was "appealing to me to be able to use nanotechnology to expand the available tools that materials scientists have, and to use that to help solve some of the problems in the world."[5]

Millie was enthusiastic. She recommended that Black review a number of articles to familiarize herself more intimately with her group's ongoing work. But looking them over, Black's heart sank. "Without a strong physics background," she later recalled, "I didn't understand 90 percent of the papers."[6] Black feared that her engineering education, which had focused on device physics, circuits, programming, and computer architecture, would prohibit her from working on the complex physics of hydrogen storage in carbon nanotubes. To her surprise, Millie didn't balk. She assured Black that no graduate student or postdoc came with all of the background they'd need in order to be a part of her group, and she offered to help make things click.[7]

Black would soon find out just how serious Millie was about bringing her up to speed. Despite considerable effort, she struggled in Millie's advanced solid-state physics course in which she'd enrolled in order to catch up. Feeling lost, Black raised her hand often in class and sought help after hours. She eventually came to the conclusion that, shy of a miracle, she was in serious danger of running aground.[8]

Millie could see Black was passionate and talented but struggling and made a move that would change Black's life forever. One morning, she announced to the class that she'd be teaching an optional recitation for the course, at what most MIT students consider an ungodly hour—8:00 a.m. "I remember her looking right at me" when she said it, Black recalls. Millie had committed to designing a semester's worth of extra classes based on material that Black needed help with, and she'd be making time to teach it week after week. "I'm very confident she did the class just for me," adds Black, who found herself the only student to regularly attend the sessions. And so, after months of one-on-one instruction, solid-state physics finally clicked: Black developed the intuition and skills she'd need to work alongside Millie, Gene, and the other group members.[9]

It was just one of countless episodes in which Millie would show her true colors as an educator who saw it as her mission to support all of her students, even when—perhaps especially when—they faced long odds of succeeding. Years later, at another particularly difficult period during Black's doctoral program, Millie's actions once again illustrated her compassion. "When my brother died, for months Millie would just skip over me at the group meetings when she asked for updates from everyone," Black says. "She knew I hadn't done any work since I was too upset still. She gave me the time to heal without the pressure of getting work done."[10]

For many of her students and collaborators, Millie was far more than just a teacher or colleague. She became a confidant, an honorary family member, and a lifelong friend. Indeed, to her sixty-some PhD students, she was both a mentor—an influential figure who offers guidance and support—and a sponsor, someone who actively champions protégés, using her voice to promote them and making time to create new opportunities for them.[11] "My mom liked to be the boss, for sure, but I think she had a strong sense of duty to encourage and to pass on things to the next generation," her son Eliot says.[12]

Of course, Millie had been going to bat for her students and others at MIT for decades before Marcie Black first set foot in her office as a wide-eyed undergraduate. One of her earliest and best-known mentees was the esteemed physicist Shirley Ann Jackson, whose subsequent remarkable career has spanned industry, government, and academia, most recently as the president of Rensselaer Polytechnic Institute. As noted in chapter 6, Jackson was one of very few Black women on MIT's campus in the 1960s—a gifted student who nevertheless struggled to find her place. Just as Millie herself had found earlier in her career, it was an era when women—and especially women of color—often found themselves swimming upstream toward acceptance as scientists and engineers.

The two met shortly after Millie joined the MIT faculty, when Jackson took one of Millie's courses on the electronic properties of materials. "I found her brilliant but also inspirational," Jackson wrote in *Physics Today* in 2017. "I admired the quality of her work, her temperament, and how she enjoyed working with young people."[13]

Millie quickly became a mentor, offering Jackson advice and an open door. "She was an incredible person, and I remember the great respect that all the other minority students had for her," Millie recalled of Jackson in Clarence Williams's *Technology and the Dream*. "Women at that time were not having the greatest time at MIT. . . . Shirley had the extra burden of being of the wrong color. It was hard for her to be accepted."[14]

Despite her struggles to fit in, Jackson excelled in her studies and thrived as a leader of MIT's burgeoning Black student population, helping to launch the MIT Black Student Union and assisting in the recruitment of more Black undergraduates to the Institute.[15] When one of her professors actively discouraged her from pursuing physics, Millie was there for her, serving as a champion. "She understood a wide range of struggles and a wide range of people," Jackson would later say of her mentor. "She certainly understood what it meant to defy those who had low expectations for women or people from less than wealthy backgrounds."[16]

In 1973 Jackson received her doctorate in nuclear physics, and in so doing she became the first African American woman to earn a PhD at MIT. But moving on from the Institute—first as a researcher at AT&T Bell Laboratories, then as a professor of physics at Rutgers University, as chair of the US Nuclear Regulatory Commission, and later as president of RPI—only strengthened her relationship with Millie as their careers followed similar paths.[17] "I'm sure I am learning more from her now than she learned from me," Millie said of her star student in *Technology and the Dream*.[18]

In a 2017 tribute, Jackson noted that Millie had become a lifelong friend and ardent supporter. "Her graceful adaptability and optimism offered me an import model as I encountered and stepped through my own unexpected windows of opportunity in industry, academia, and government," Jackson said. "I am forever grateful to Millie Dresselhaus for her friendship, for her warmth, and for her example."[19]

HOT AND COLD: REVIVING THERMOELECTRICS

What if clothes could automatically cool a wearer down, eliminating the need for energy-intensive air conditioning? What if we could reduce our energy consumption by turning the heat expelled by vehicle engines into electricity? What if we could eliminate wasteful overheating in electronic devices by building components that would cool themselves?

Coincident with the confirmation of single- and multi-walled carbon nanotubes, the 1990s was a transformative decade for Millie, as the expanding science of nanotubes provided myriad mysteries to solve and potential applications to develop. These years also became a focal point of Millie's career thanks to her preeminent role in the revival of a field in which scientists and engineers have attempted to harness a principle of physics known as the thermoelectric effect for the development of new, energy-efficient devices.

First discovered in the late eighteenth century by Italian scientist Alissandro Volta and further described in the next century by Estonian physicist Thomas Seebeck, Danish physicist Hans Ørsted, French physicist Jean Peltier, Irish physicist William Thomson (Lord Kelvin), and others, the thermoelectric effect comprises two separate actions: the tendency of a temperature difference between the two sides of a material to convert spontaneously into an electric current and the reverse scenario, in which applying a current through a material can generate heating or cooling without any moving parts (figure 9.1).[20]

Researchers began working in earnest to engineer products that would take advantage of the thermoelectric effect in the

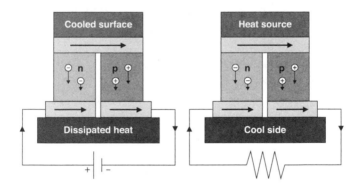

FIGURE 9.1

Thermoelectric devices create a voltage when the two sides of a material are different temperatures. Here, thermoelectric circuits composed of two semiconducting materials ("n" and "p") are configured as a thermoelectric cooler (left) and a thermoelectric generator (right).

1950s and 1960s. Companies including RCA, 3M, and Texas Instruments dedicated some of their efforts to thermoelectric devices including power generators for spacecraft and coolers for infrared sensors. But by the 1970s, the field collectively hit a wall, and research in thermoelectrics stalled.[21] According to John G. Stockholm, an engineer who specializes in thermoelectrics, "The only thermoelectric materials, since the late 1950's, were bismuth telluride, lead telluride, and, to a lesser extent, silicon-germanium alloys. . . . No universities were interested in new materials for thermoelectric application, and no funding was available."[22]

What happened? The central physics and engineering challenge that has plagued thermoelectrics is that its key materials need to be excellent electrical conductors while limiting heat conduction. But in most materials, enhancing electrical conduction necessarily increases heat conduction, and vice versa, as each charge carries a lot of heat. It proved quite difficult to separate current from heat, and even when engineers were able to work around this, irreversible heat loss was inevitable, making efficiency very low. As a result, thermoelectrics simply couldn't compete with traditional means of heating and cooling.[23]

The dawn of the nanoage helped scientists and engineers revisit some of these problems, and Millie was a vital spark in the revival of the field. Around 1990, both the French Navy and US Navy approached her about developing thermoelectrics for submarine applications. They were particularly interested in devising ways for subs to generate electricity without any discernible effects—no noise, no bubbles, and no exhaust—to help them traverse international waters undetected.[24] "They asked me if I had any ideas," Millie noted in a 2007 oral history, "so I said, well, we have all these new materials around now, nanothings. That could be interesting."[25]

Indeed it was. When MIT graduate student Lyndon Hicks approached Millie for a research topic around the same time, she suggested they look into thermoelectrics together. Several of their papers, published in 1993, helped to jump-start the subfield of nanothermoelectricity, which focuses on using quantum mechanical effects in nanometer-sized molecules to improve electrical conductivity while reducing heat loss. They looked specifically at "low-dimensional" systems, meaning they made predictions on two-dimensional nanoscale lattices that were engineered to interrupt the flow of heat, as well as one-dimensional nanowires of varying types. They would work by, for example, adding tiny particles to an existing nanostructure that would physically block the flow of heat while allowing electricity to flow.[26]

The results were highly influential. Among other things, Millie and her colleagues and students showed that nanowires and superlattices had strong potential to be used for more efficient thermoelectric devices.[27] Significant progress in materials performance has since been made thanks to continued research in thermoelectrics, but the field has not yet realized its full potential. In a 2013 article looking on twenty years since Millie's first papers with Hicks, she and several colleagues considered why highly efficient thermoelectric materials still hadn't arrived. For starters, they noted that government funding of this line of research had been weak. Manufacturing viable materials at the right scale had also "presented technical and cost challenges."[28] But researchers continue to work toward a future in which thermoelectrics would facilitate new applications—such as cooling systems for microchips, car engines that repurpose the heat they generate, and electricity created from heat that a power plant emits—that might help to mitigate climate change.[29]

For many years, Millie collaborated with MIT professor of mechanical engineering Gang Chen to improve the efficiency of thermoelectric materials for such purposes. Chen began his association with Millie as a young professor at Duke University; when he moved to UCLA, they would discuss thermoelectrics after her meetings at Caltech, where she served on the board of trustees. Chen would also shuttle her to the Los Angeles airport, where she'd inevitably catch a red-eye back to Boston.[30] "Her luggage was always so heavy!" he says. Much of the heft came from papers, some of which, in the days before the Internet, she would ask friends and associates to fax to Gene and others so they'd arrive back in Cambridge before she did.[31]

After Chen eventually settled at MIT, he and Millie saw each other often. She was a coinventor with Chen on five US patents, mostly for methods of creating nanostructured thermoelectric materials.[32] "I think Millie's biggest contribution to the field is getting people to rethink," he says. "Before her, there was not much funding, and the quality of work was low. Her work really opened people's eyes to say, 'Oh, there are other ways!'"[33]

Though they were in different departments, Chen and Millie were among the first to arrive on campus each day and would routinely see each other there at 6:00 a.m. or earlier. Throughout their association Millie served as a mentor, particularly when Chen juggled research with the responsibilities of a department head from 2013 to 2018. He also enjoyed visits to Gene and Millie's house in Arlington and walks in nearby parks.[34]

"At MIT, there are many Jedi knights, but Millie stands out as our Yoda," Chen once wrote in a letter to *MIT Technology Review*. "Students and faculty members alike seek out her advice and opinions about a wide range of problems, in both their research and personal lives. Warm and open, she is always receptive, ready to work, and willing to help."[35]

DR. DRESSELHAUS GOES TO WASHINGTON

Millie's eagerness to help others in her professional orb extended well beyond the walls of MIT and far into the public sphere. In the 1990s and 2000s, she expanded her already considerable science service by engaging in some of the most influential and prestigious national service projects that a scientist or engineer can undertake.

For Millie, such projects took on deep personal meaning: She felt strongly that since she'd been on the receiving end of countless opportunities to advance in education and in science, paid for by government and private organizations, beginning with her earliest days at the Greenwich House School of Music in New York, she was obliged to give back.[36] This was an attitude she learned at her alma mater, Hunter College, where tuition was nearly free—just $5 per semester, or about $50 today.[37] In an interview for Martha Cotter and Mary Hartman's *Talking Leadership: Conversations with Powerful Women*, Millie remarked, "One message I recall very clearly, and which has in fact been a light of my career, was the following: 'This free education is not free. We expect you to pay this back in service to society over your lifetime.' I think that's a very good exchange for all concerned!"[38]

In February 1996, Millie was announced as president-elect of the American Association for the Advancement of Science (AAAS), the world's largest science organization with over 140,000 members at the time. She served as vice president the following year and assumed the presidency in 1998, only the ninth female president of the AAAS since its founding in 1848. The first, in 1971, had been Mina Rees, the mathematician who spoke at Millie's Hunter College graduation and met Millie afterward, encouraging her to continue with her math and science studies.[39]

In her year as president, Millie focused on such key areas as educating scientists on budget issues so they could maximize sometimes scarce funding; improving public science communication among members; boosting participation in AAAS activities from members representing a broad range of science disciplines; and attracting and encouraging the next

generation of scientists. But perhaps her proudest accomplishment was recruiting President Bill Clinton to speak at the
1998 AAAS national conference marking the 150th anniversary of the association. She maintained that the experience
helped Clinton transition from a position of tepid interest to
one of strong support for science and technology in his later
presidency.[40]

In his keynote, Clinton spoke on many issues, including
cancer research and the importance of reducing teen smoking,
the Human Genome Project and ethics surrounding human
cloning, the development of alternative energy sources, science education and diversity, and the impending information age (with some prescient predictions about the ease with
which society would soon have access to "every book ever
written, every painting ever painted, every symphony ever
composed"). In a rousing closing, Clinton made a hopeful
forecast for the future of science and society:

> Your bicentennial meeting can convene in a world where climatic
> disruption has been halted; where wars on cancer and AIDS have
> long since been won; where humanity is safe from the destructive
> force of chemical and biological weapons, wielded by rogue states
> or conscience-less terrorists and drug runners; where our noble
> career of science is pursued and then advanced by children of every
> race and background and where the benefits of science are broadly
> shared in countries both rich and poor. That is what I pray it will
> be like, 50 years from now, when my successors stand here before
> your successors and assess how well we did with our time.[41]

Two years later, in January 2000, Clinton announced a
major new government program, the National Nanotechnology Initiative. It was a wide-ranging policy focused on enhancing basic science related to matter at the nanoscale, developing
new applications across many industries, and supporting and
sustaining a workforce focused on nanoscale science and engineering. The program would fund some $12 billion in research

and development in its first decade, a massive government effort that trailed only NASA in terms of civilian science and technology spending.[42]

That spring, Clinton also honored Millie by nominating her to an important position within his administration. It took a few months to clear Senate confirmation, but a June 2000 article by Irwin Goodwin in *Physics Today* gracefully summarized public sentiment about Millie's new post:

> At the dawn of his presidency in 1801, Thomas Jefferson wrote: "There is nothing I am so anxious about as good nominations, conscious that the merit as well as the reputation of an administration depends as much on that as on its measures." Jefferson's standards were attained many times over the centuries and once again were met on 13 April when President Clinton nominated Mildred S. Dresselhaus, a prominent solid-state physicist at MIT, to serve as the next director of the Department of Energy's Office of Science.[43]

Millie knew it would be a brief engagement: Clinton was in the final year of his presidency, and it was (and remains) unusual for individuals to keep their posts after the installment of a new presidential administration. But she made the most of her time in Washington, taking a leave of absence from her duties at MIT to oversee one of the largest US sponsors of basic science, with a budget, at the time, of $2.8 billion. In true Millie fashion, however, she returned to MIT every weekend to support her students.[44]

During her tenure at the Department of Energy, Millie managed a number of laboratories and activities related to government-sponsored energy research and development, as well as science education. Among other efforts, she worked to reverse declining budgets for the physical sciences, spent time developing a system for evaluating which projects to fund, and served as a strong supporter of both young scientists and scientists-to-be. A September 2000 visit to Fermilab, a high-energy particle physics laboratory named after Millie's graduate school mentor, illustrated her dual role as pragmatic truth-teller and inspiring leader.[45] In summarizing the trip, a Fermilab newsletter noted her assertion that "there should be no whining and moaning" about budget woes facing high-energy physics at the time, while also quoting a graduate student on Millie's support for young people at the lab: "The meeting with Dr. Dresselhaus was like doing group therapy. . . . I feel really happy to come in to work today."[46]

On December 12, 2000, the US Supreme Court declared George W. Bush the forty-third president of the United States after a contentious recount dispute in the November presidential election. With the looming change in administration, Millie took leave of our nation's capital and returned to Massachusetts and to MIT.

GETTING TO KNOW YOU

For the majority of academics, being distinguished with a single honorary doctorate from an institution of higher education is considered the height of acknowledgment for a lifetime of contributions to a field. But for pathbreaking scholars like Millie, whose contributions know no bounds, honorary degrees tend to multiply quickly. Overall, Millie received nearly forty of these recognitions; from 2001 to 2010, she averaged one per year, and the pace picked up slightly in subsequent years. But far from simply arriving for the ceremony and heading back home, Millie used such opportunities to connect with scientists and other academics at the host college or institute and to directly encourage the next generation of physicists, chemists, materials scientists, and engineers—whether through a specially crafted speech, an in-person meeting, or the opportunity such an award might confer in terms of helping to select subsequent honorees.[47]

It was around this time that Millie also took on two signature marks. (Her original identifying mark, a perpetually
braided updo, began in the late 1950s, just before she joined
Lincoln Laboratory. Her family tells me it was inspired by an
Austrian hairstyle that ensured it was always neat and out of
the way in the lab.) One new mark was her Scandinavian-style
knit sweaters, for which she came to be known in the last
decade of her life. She purchased her first Norwegian sweater
in 2004, and they soon became a unique identifier: At conferences, at the airport, or at awards ceremonies, you could
spot Millie a mile away thanks to her sweaters. The brighter
of the two cardigans she wore most often was an intricately
woven cardinal red with wine and black accents and silver
buckles; the other cardigan I consider her snow-owl sweater, a
black-on-white affair featuring a central floral motif and dot-
ted arms.[48]

Her second new mark was the nickname for which she
became known around the world. The exact origins of "Queen
of Carbon" (or "Queen of Carbon Science") are somewhat
shrouded in mystery, but it seems to have solidified as a recognized moniker in the mid-2000s. Millie herself once claimed
that she'd earned it from journalists on Swedish television.
But family members say it's likely she'd heard it before, as a
nickname from students and others she'd worked with. "At
first, Millie really disliked the title and rejected it," says daughter Marianne.[49] Granddaughter Shoshi Dresselhaus-Cooper
adds: "Millie didn't want to be treated like royalty. She always
wanted to just enjoy the science without seeming unapproach-
able."[50] At some point, however, Marianne pointed out that
the nickname was actually quite appropriate: Millie's Hebrew
name was Malka Sheindel, which loosely translates to "queen
treasure." After that, Marianne says, Millie started to accept
it a lot more. According to granddaughter Leora Dresselhaus-
Marais, doing so also allowed her to more actively engage a

new generation of women and girls who might benefit from her example as a role model in science.[51]

"CYCLES IN SCIENCE": BACK TO GRAPHENE

The new millennium saw Millie, Gene, their students, and close colleagues continuing full-speed on multiple fronts. A heavy stream of coauthored papers and invited talks in the early 2000s provide a window into Millie's areas of academic interest during this time: They focused mainly on the science of carbon nanotubes, Raman spectroscopy of nanoscale carbons, bismuth nanowires for thermoelectrics, and science policy and practice. In 2003, she took on another important national leadership role when she was elected chair of the American Institute of Physics governing board, the first—and so far only—woman to achieve this position. That same year, she also chaired a Department of Energy workshop on research needs for the coming hydrogen economy.[52]

A significant shift took place in the mid-2000s, after a team of physicists from the University of Manchester in England achieved a breakthrough that had been more than a century in the making: developing a relatively simple way to isolate and study monolayer, or single-atom-thick, sheets of graphene.

The primary ingredient of graphite had been known theoretically to scientists and engineers since the mid-twentieth century, and sightings of graphene were made even earlier. In 1859, British chemist Benjamin Brodie observed what we now know to be graphene oxide—graphene flakes covered with additional organic materials. Nearly ninety years later, in 1947, Canadian physicist Phillip Wallace developed and published important theoretical work describing graphene's electronic structure. His research, however, was barely recognized within the larger physics community, which was not particularly interested in nanoscale carbon at the time.[53]

The pace of graphene discovery picked up in the 1960s. As we saw in chapter 5, high-quality synthetic graphite came along just as Millie began her career in carbon science. Back then, she and the smattering of colleagues she had in this nascent field began using a new technique for removing flakes off their graphite samples: sticking it to cellophane tape. The idea was to prime the graphite for investigation into any number of properties, but notably, Millie did not attempt to isolate the individual graphene layers that stuck onto the tape; they were instead flung into the trash.[54] "At that time, graphene was considered as an intellectual prototype model system . . . and as an esoteric research topic that attracted little interest among the research community," she wrote in a 2011 article.[55]

In 1962, German chemist Hanns-Peter Boehm and three colleagues reported the synthesis of few-layer and monolayer graphene, but this too went unappreciated by the physics community at large. Among other things, the prevailing opinion of the day was that (virtually) two-dimensional graphene couldn't possibly be stable on its own—neither, for that matter, could curved graphene-based structures like buckyballs or carbon nanotubes.[56] "It was controversial; there were many people who were skeptical," Millie told *MIT Tech Talk* in 2009.[57] Tools used to probe graphene samples were also quite unsophisticated compared to those available today; pure graphene is a nearly transparent material, so it would have been difficult to know for sure how many layers one might have isolated from a graphite sample.[58]

Novel ways to prepare graphene were developed in the early 2000s by Dutch physicist Walt de Heer, Japanese physicist Toshiaki Enoki, Korean American physicist Philip Kim, and others. But it wasn't until 2004 that a veritable gold rush in graphene research truly began. That was the year in which a team of eight physicists, led by Andre Geim and Konstantin Novoselov at the University of Manchester, first described in the journal *Science* their now-famous process for preparing few-layer and monolayer graphene. Central to their method was cellophane tape, just as Millie and others had used decades earlier. But Geim and his colleagues—including Irina Grigorieva, an expert in two-dimensional materials and Geim's wife—had developed a way to isolate bits of graphene just a single atom thick and, critically, confirm they'd done so.[59]

The beauty of their experiment lay in the setup. After lifting graphene flakes off a graphite sample with tape, they stuck the resulting carbon onto a silicon oxide platform. Doing this affects the perceived color of the silicon oxide, which means that a scientist can determine the thickness of her graphene based on the color of the silicon it's sitting on. Anything thinner than ten layers is considered few-layer graphene (as opposed to a larger "thin film" of graphite) and appears pink under a light microscope; anything that's just one layer thick—a graphene monolayer—is very light pink.[60]

The 2004 paper made instant waves not only for its clear demonstration of graphene isolation but for the confirmation of new physics, especially the material's behavior as a high-quality electrical conductor despite being only one atom thick. Physicists who had already been interested in carbon nanostructures—as well as many who became intrigued enough to refocus their attention—dove into this newly promising material. As Millie often described in talks around the world, the number of graphene-related research articles skyrocketed. Some of these new works focused on ways in which to synthesize high-quality graphene for use in various applications, while others focused on new understandings of graphene physics.[61]

Based on decades of theoretical work on graphite and its individual layers, scientists and engineers knew that graphene applications, should the material prove stable enough to work with, had the potential to be game-changing in a number of

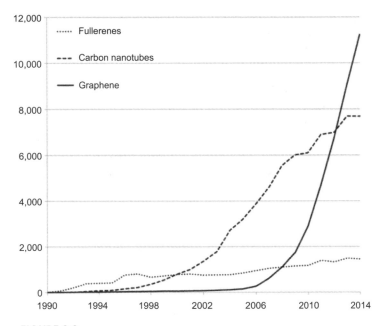

FIGURE 9.2

A quantitative look at scholarly citations related to fullerenes, carbon nanotubes, and graphene attests to the dramatic rise in graphene research since its isolation in 2004. Credit: Reprinted from E. M. Pellegrino, L. Cerruti, and E. M. Ghibaudi, "Realizing the Promise— The Development of Research on Carbon Nanotubes," *Chemistry: A European Journal* 22, no. 13 (2016): 4330, https://doi.org/10.1002 /chem.201503988.

important fields, from next-generation transistors and flexible electronics to quantum computers to improved composite materials to biological films for various medical applications. "It was as if science fiction had become reality," Youngjoon Gil, executive vice president of the Samsung Advanced Institute of Technology, told the *New Yorker* in 2014.[62]

To be sure, many researchers also were excited about potential applications they hadn't even dreamed up yet. "It's the most extreme material you can think of," MIT professor Tomás Palacios told *MIT News* in 2009.[63] At a single atom thick, graphene is the thinnest possible material and extremely lightweight. By virtue of the strong covalent bonds connecting its carbon atoms, it is, like its nanotube cousins, immensely tough—far stronger than steel. Graphene is also a superior conductor of electricity, with electrons coursing through its carbon-atom lattice at about 1/100 the speed of light, making it an excellent candidate to rival silicon as the basis for new, more efficient devices.[64] Who wouldn't want to spend time making discoveries with a substance like that?

For her part, while Millie continued important work on carbon nanotubes and thermoelectrics, beginning in the mid-2000s she also refocused on graphene for what would be the final stage of her career. Among many others in nanoscience, her students and postdocs' curiosities were piqued by this "new" material, and she exuded excitement along with them. Millie also welcomed the opportunity to improve on some of the ideas she'd first worked out earlier in her career. She noted in a 2007 interview that a number of researchers had in fact been redoing some of the experiments she had performed decades earlier, this time with added background knowledge of and, of course, better equipment. "It's like cycles in science," she said.[65]

Over the next decade, Millie coauthored dozens of papers on graphene. Her work focused on everything from general spectroscopic studies of the material, including its electrical and thermal characteristics, to detailed properties of graphene nanoribbons, strips of graphene no wider than about 50 nanometers; to what graphene does when it's damaged or "doped" with other elements; to the morphology and behavior of graphene at its edges, both of which are affected by the alignment of its hexagons. She was also interested in finding ways to synthesize relatively large sheets of high-quality graphene

(and other two-dimensional materials), which would facilitate manufacturing of the materials and spur further
research.[66] "It's not going to be Scotch tape forever," she said
in 2009.[67]

NEW BEGINNINGS

In 2004 Millie gained a new colleague—and friend—in Jing
Kong, a Chinese researcher with a background in carbon
nanotubes who had come to the United States to further her
chemistry education. Millie welcomed Kong, a bright, young
assistant professor newly hired by MIT, as only she could. Not
only did she share her office space when Kong hadn't yet one
of her own, she also offered Kong the opportunity to work
with her students—ensuring that she'd immediately become
absorbed into Millie's world. "I think," Kong says with reverence, "Millie volunteered to be my mentor."[68]

Kong would eventually settle just down the hall from Millie, whose guidance proved to be formative for the young
professor. The two would become longtime research partners;
over a dozen or so years, they shared many students and coauthored more than sixty research papers together.[69]

One of the first topics Millie suggested that Kong explore
was graphene synthesis. After reading the fateful 2004 *Science*
paper by Geim, Novoselov, and colleagues, Millie asked Kong
if she might be interested in working out ways to systematically produce graphene. "That really got me thinking," Kong
recalls. "Graphene and carbon nanotubes [are] very similar,
and we use a similar synthesis strategy to obtain graphene.
That started my research in 2D materials."[70]

Millie and Kong collaborated on research topics ranging
from the properties of graphene created by chemical vapor
deposition to defects in carbon nanotubes to ways to safely
transfer graphene sheets onto various surfaces. Their students

often sat in the same group meetings, and they benefited
from Millie's myriad international colleagues, who would
take advantage of any opportunity to visit MIT and perform
research with Millie and Gene.[71]

But it wasn't all work between Kong and Millie; the two
shared a great deal of time together outside MIT, especially
in the last years of Millie's life. For most of their overlapping
years on the MIT campus, Gene drove Millie in every day from
their home in Arlington. But he developed health problems
in the early 2010s and officially retired in 2013, prompting a
shift in Millie's commute. Thereafter, she had help from other
drivers, including Kong, who often drove Millie home.[72]

Just like Millie and her own mentor, Enrico Fermi, back
when the two would walk together to the University of Chicago physics department, Millie and Kong bonded over a
shared commute and a mutual love of science. "Millie told me
a lot of stories," Kong notes. "She always tried to make good
use of every opportunity: For every [car ride] she would come
with a topic to discuss—work, life, everything." Negotiating
the streets of Cambridge and Arlington, the two often talked
shop, including comparing notes on students and coming up
with strategies for those who might be struggling. But they
just as often discussed their families—in particular, the lives
of their children and Millie's five grandchildren. "She always
respected her kids' interests," Kong says. "As a mom, that was
a big [lesson] for me; how not to impose my own thoughts on
them, but encourage them."[73]

Kong is now a full professor in the MIT Department of Electrical Engineering and Computer Science, and a noted expert
in two-dimensional materials. As a visual testament to her
impact on the field, her office's bookcases are topped with a
dozen or so champagne bottles marking the successful thesis defense of her doctoral students. She beams when talking
about Millie: "The way she liked to mentor students and how

much she enjoyed what she is doing," Kong affirms, "really set such a good example for me."[74]

On the morning of October 5, 2010, carbon nanoscience was once again in the limelight when Andre Geim and Konstantin Novoselov were awarded the Nobel Prize in Physics "for groundbreaking experiments regarding the two-dimensional material graphene."[75] It bears noting that while their 2004 and 2005 publications were certainly transformative in terms of igniting interest in graphene, the prize didn't just recognize one discovery; rather, it served as a reflection of their continued advances at the leading edge of a newly emerging science. "The Geim-Novoselov team maintained their leadership position in this fast moving field by continually generating new, exciting results," Millie wrote with Paulo Araujo, a former postdoc in the Dresselhaus group, in a brief history of graphene following the Nobel announcement. "It was the high quality and impact of these publications and public presentations that made an indelible mark on the graphene field and on science more generally."[76]

Starting with the superconductivity physics prize in 1972, this marked the third Nobel awarded in Millie's lifetime that she had some connection to. Her role with the graphene prize was particularly significant, and both of the winners thanked her directly in their Nobel speeches: Kostya Novoselov listed her among a dozen or so physicists who had influenced his research, and Andre Geim mentioned her twice—once as a source of inspiration through her foundational work on graphite intercalation compounds, and second as a source of insights on theoretical work relating to carbon nanotubes with Gene, Saito, and Fujita. She was also invited to attend the Nobel ceremonies as Geim's honored guest.[77]

Was Millie disappointed that in all her years of nanocarbon leadership, including a number of widely influential individual discoveries, she had never been recognized with a Nobel Prize herself? The simple answer is: not at all. For one thing, as we'll see in the next chapter, she was highly decorated with just about every other top prize in physics and in science. And while the popular media eventually made a point of mentioning her in their annual Nobel season articles as a deserving candidate, it was not something she spent any appreciable time thinking about.

Laura Doughty, Millie's longtime assistant, stresses that seeing research advance was always Millie's biggest reward: "She didn't feel the need for recognition, she didn't feel the need for credit. She wanted to see science furthered."[78]

Millie even admitted in a 2002 interview to feeling some guilt after receiving so many awards for work done in collaboration with others. "I'm well known in my field," she stated, "and that's enough."[79]

10 AN INDELIBLE LEGACY

On the first Saturday in December 2010, Millie Dresselhaus and 250 of her closest colleagues, students, friends, and associates came together to party. Ever-ebullient Millie had turned eighty just a few weeks prior, and a special symposium in her honor had been organized as a way for all who knew and loved her to toast an incredible career and life to that point.[1]

Millie had become an emerita professor three years earlier, exactly four decades after beginning her storied career on MIT's campus as a young visiting professor of electrical engineering. Including her tenure at Lincoln Laboratory, her nearly half a century at the Institute to that point spanned a third of the school's entire history.[2]

But the reality is that Millie had a difficult time with the concept of slowing down—and she never really did. While she stated on multiple occasions that she was planning to "someday" retire, there never seemed to be a right time.[3] "Let me tell you that being 80 years old isn't so bad," she quipped at her birthday bash. "As long as you have science on your mind and fun things to do, you really don't feel the age at all."[4]

Indeed, it took only a fleeting glance at her office in Building 13 to know that even in the twilight of her career, Millie still had loads of science on her mind. Her room was chockablock with mountains of papers and books and lined with ball-and-stick chemical models as well as mementos from her countless collaborations and excursions around the planet. You got the sense that she'd have hated the popular Marie Kondo process of decluttering, if only because, for her, so much of what filled her workspace "sparked joy."

Visitors to Millie's office were welcomed into a small, central seating area—a smart arrangement, lest they not be able to see the Institute Professor as she spoke over piles of stuff on her desk. "This is my little hole," she once told the *Boston Globe*, "my little sea of paper that needs to be cleaned up."[5] She wasn't, of course, the first physicist to make her professional home in a cluttered office, and she won't be the last; her own MIT colleague, theoretical physicist Alan Guth, once won a contest for messiest office in Boston.[6] But Millie's longtime assistant Laura Doughty confirms that her boss's workspace was often out of control. In fact, when President Barack Obama visited her building in 2009, the office was flagged as a potential fire hazard.[7] "We had the fire marshals in there," Doughty says.[8]

Millie was not an early adopter of communication technology. In her later years, she claimed her iPad, which she learned to use in 2012 with help from her assistant Read Schusky, was a favorite device. But she didn't really use computers for communications until the early 2000s. Her first cell phone, a Blackberry, was given to her when she served at the Department of Energy in the Clinton administration, and using it marked the first time Millie systematically wrote and sent her own email.[9] "That was revolutionary!" Doughty exclaims.[10]

Until that point, Millie relied heavily on the phone and especially the fax machine, which contributed to the piles in her office. Tales of how she'd tote suitcases of documents on her frequent travels around the globe, marking them up with red Pilot Razor Point pens as she waited for connections and finding a way to send them off as soon as possible after landing, are legion.[11] But even after she started using email and PDFs more frequently, she still preferred to mark work up by longhand. "She did a huge amount of collaborating on papers, and that was always hard-copy—100 percent," Schusky says.[12]

SOMETHING SACRED

In a late 1990s interview with *MIT Tech Talk*, Millie revealed that the secret to her long success was retiring for the night by 11:30 and getting to MIT at 5:30 or 6:00 every morning. Doughty recounts that until her latest years, Millie did quite well with very little sleep—and that she'd often sneak in naps at her desk, refreshing her for many more hours of work.[13]

But another key to her success was making sure to have a creative outlet beyond her academic and service work. For Millie, that was always music. It began with her earliest music lessons in New York City, continued with participation in musical opportunities at Cambridge University and through the Arthur von Hippel quartet she played with at lab parties and other events, and flourished through the ad hoc chamber groupings she'd convene, often in her home, bringing hundreds of centuries-old harmonies to life with her nimble fingertips.[14]

"Millie didn't watch TV, she didn't watch movies, she didn't read books," Doughty says. Instead, her principal avocation was playing violin and viola, although she very much enjoyed hiking and cooking as well. Doughty, as both the manager of her calendar and a professional singer, had an especially close view of Millie's commitment to music. Doughty saw firsthand how Millie made time for music nearly every day when she wasn't traveling—and how she shared her experiences with family, students, friends, and many others who participated in informal music nights at their home or elsewhere. "The evening calendar, which Gene arranged, was just as important as the daytime calendar," she says. "The music calendar was sacred."[15]

Aviva Brecher, a former student who went on to a multi-faceted career in planetary science, nuclear science, transportation, and more, shared fond memories of Millie's musical world at a 2017 tribute event: "I remember one of the early dinners that I, an undergraduate, and my husband Ken, an MIT graduate student, were invited to. . . . We had a lively dinner and, afterwards, all four Dresselhaus kids played us a string quartet—with Gene turning the pages! It was very welcoming and truly amazing."[16]

Brecher also recalled interludes in MIT's Cheney Room, where the Women's Forum began in 1972. "I played the piano there with Millie . . . who brought her violin so we could play Tchaikovsky," she noted in a 2017 oral history. "Millie brought the civilizing effect of music, by playing music before every [electrical engineering] seminar. This was unheard of at MIT."[17]

According to Laura Doughty, who sang in chamber groups at the Dresselhaus home on multiple occasions, Millie's commitment to music was like her attitude toward her constantly evolving career. "They never stopped," she says. "I'd want everything to be perfect, but . . . they had no patience for that. You played it through, and if you lost your place or missed some notes, you just got back on. Millie cultivated that, and that's how, I think, she was in life: If you lost your place or if you got confused, you just get back on. You don't stop, you don't complain, you don't make excuses."[18]

AN EXTENDED FAMILY

For Millie, the precocious kid from Brooklyn who'd been soaking up the whys of the world since her childhood, an intense intellectual curiosity never waned. But it wasn't just the pursuit of new scientific understanding that drove Millie to stick with her investigations of carbon, bismuth, and other materials well into her eighties. The camaraderie and friendships—with staff and faculty colleagues, students and postdocs, and collaborators around the world—were equally important. "MIT has always been my extended family. And my students are like my children," she stated in 2007.[19]

Millie and Gene opened their home regularly for their associates. In addition to music nights, they often invited students, colleagues, and others to fill their Arlington abode with laughter, food, and conversation. In particular, many students have described being invited to Thanksgiving dinners, prepared and hosted annually by Millie and Gene and open to anyone in the group who had nowhere else to go.[20] "It was not like Millie picked up a few stragglers—she took in most of the lab," says Mario Hofmann, an associate professor of physics at the National Taiwan University who was a PhD student in the Dresselhaus group from 2005 to 2011.[21]

The Dresselhaus professional family was far from confined to MIT, of course. From her earliest years as a professor, Millie's collaborations were global (figure 10.1). Laura Doughty recalls Millie planning her days based on the time zones of her colleagues—cleverly choreographing outgoing messages in an intricate digital dance so that recipients had the most possible time to twist and turn them around before their day ended. And for most of her career, wherever she trekked, Millie would know someone—a former student, a research associate, or some group that would roll out the red carpet, making arrangements for her visit and otherwise treating her like a favorite aunt or grandmother.[22] It was, she said in a 2007 oral history, "the benefit of science—that science is the universal language. We know each other, and people care about other people."[23]

"Millie's students were a continuous thread of contributions and personal support," affirmed former graduate student

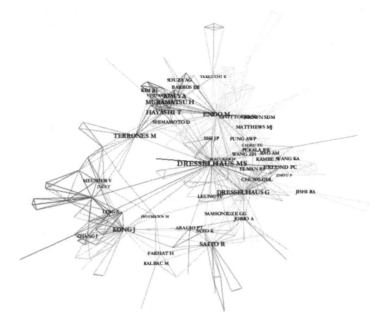

FIGURE 10.1
Physicist Mario Hofmann, a former PhD student in the Dresselhaus group, was inspired to quantify Millie's academic family. Using data from her nearly 1,700 research publications, he concocted a "family tree" brimming with branches—4,872, according to Hofmann. Only two of her eight most common collaborators, Jing Kong and Gene Dresselhaus, were MIT colleagues; the rest hailed from around the globe.

Mario Vecchi, now chief technology officer at the Public Broadcasting Service, at a 2017 tribute event. "At the core of Millie's success was the person that she was, and the values that she brought to her work and her personal life. . . . Millie and Gene created the most inspiring environment for their children and for all their friends."[24]

From time to time, it was not unusual for Millie to throw her considerable professional weight behind a former student or postdoc who needed it. For Calvin Lowe, one bright MIT scholar who had struggled both academically and socially in the early 1980s but nevertheless seemed certain to shine given the right opportunity, Millie helped to secure a position for him as an assistant professor at a university where another of her former students was already on the faculty and able to offer direct support.[25] "She never gave up on Calvin, and he eventually became a vice president for research at a historical Black university, and then president of a university," says longtime MIT colleague Clarence Williams. "Mildred Dresselhaus played the most important role in his career."[26]

More recently, Millie played a critical role in advocating for and then helping a former visiting student and his brother land new positions after they were controversially dismissed in 2009 over disagreements with their former employer in Mexico.[27] "Millie made our transition to the US . . . very smooth," said Mauricio Terrones, now a physicist at the Pennsylvania State University, in 2017. "She helped us, and she wrote very nice letters of recommendation. That's why we're here."[28]

For former student Mario Hofmann, Millie's willingness to help out a stranger altered his life enormously. "My first internship position had not materialized, and one night my advisor picked up the phone and simply called Millie, who he had literally seen once in a conference," Hofmann says. "He introduced himself and described my situation, and without a moment's hesitation Millie agreed to go through the hassle and financial burden of inviting me to MIT. I was close to hyperventilating because she had literally written the book [related] to our nanocarbon work."[29]

Time and again, Millie's support—a recommendation letter here, a back-channel endorsement there—vaulted students, colleagues, and others into key positions that would set them on their life's path. This only enhanced Millie's reputation as a den mother to those she worked closely with, and her support created lifelong appreciation, even among individuals with whom she was only loosely affiliated.

Wanting to create such a familial atmosphere for her associates did, however, have its limits. By the final decade of her career, the volume of recommendations, requests, and favors Millie was asked for became nearly impossible to manage. "The numbers were just so huge and so constant, it was like a river," Laura Doughty says. "The fact was Millie rarely said no. She'd always say, 'Hm. I could do that!' Her brain was always in problem-solving mode, so if someone asked her to do something, even before thinking, 'Do I want to do this or not?' she'd be thinking, 'Well, how can I do it?'"[30]

Of course, there were professional requests and then there was family time. As Millie's family grew to include five grandchildren, she made an effort to spend as much time as she could visiting them. To make this possible, she often coordinated family visits with professional engagements. She also planned for each grandchild to enjoy a special "Millie trip" to some far-flung destination.[31]

"I only realized the extent of Millie's fame when I traveled with her to England and China in the summer of 2011," says granddaughter Elizabeth Dresselhaus, a PhD student in theoretical condensed-matter physics at the University of California at Berkeley. "People on the other side of the world recognized her name and her work [and] wanted pictures with us. It was incredibly surreal."[32]

"Something I think people don't always realize about Millie is that she had a spirit of adventure," adds granddaughter Clara Dresselhaus, an undergraduate majoring in applied mathematics and psychology at the University of California at Davis whose "Millie trip" was to Oxford, England. "We had an interesting experience punting on the Thames River, and it ended up with my falling off the boat and into the water. When we got back to shore, I had the elegance of a wet cat, and Millie was sitting . . . with a big smile, cackling."[33]

HONORING A PIONEER

A petite octogenarian in an elephant-gray fitted blazer, coal-colored slacks, and a crown-braided silver updo sits quietly among some of the most famous Americans of the twenty-first century. To her left is a bespectacled actor, often described as the "best of her generation," who over a forty-five-year career has garnered three Academy Awards and a record twenty-one Oscar nominations for her work in films such as *Kramer vs. Kramer*, *Sophie's Choice*, and *The Iron Lady*. Farther down, in the row ahead, sits one of the most influential musicians of all time, a man who has won twenty-five Grammy Awards for songs including "Superstition," "You Are the Sunshine of My Life," and "I Just Called to Say I Love You."[34]

When it comes time to stand, the lone scientist among the day's honorees does so with grace and a broad, beaming grin. At the presentation of the 2014 Presidential Medals of Freedom—which, along with Congressional Gold Medals are the highest civilian awards given in the United States—the pioneering scientist and engineer seems right at home as a luminary among famous faces of journalism, public service, and the arts.

"Her influence is all around us," Obama says of Millie in his introductory remarks, "in the cars we drive, the energy we generate, the electronic devices that power our lives." When her name is called to receive her medal, she stands, scanning the room as her citation is read: "one of the most distinguished physicists, materials scientists, and electrical engineers of her generation" and "a leader and mentor" whose "example is a testament to what we can achieve when we summon the courage to follow our curiosity and our dreams."[35]

Millie Dresselhaus was the acclaimed recipient of countless honors throughout her career. But in the final decade of her life, awards for half a century of research and leadership multiplied rapidly, and her public persona grew to reach an entirely new plane. "We had to work on her to learn to enjoy the recognition," former assistant Laura Doughty says, adding, "Millie understood the desperate need for the public to grasp the importance of science. By giving up time for photo shoots and interviews, she recognized that she contributed to that endeavor."[36]

Among the tributes that helped shepherd a new appreciation for the magnitude of her own achievements and the power of her influence was the L'Oréal-UNESCO for Women in Science Award, a global award that recognizes outstanding achievement in life and materials science. "How can there be anybody better?" asked physicist Peter Eklund, one of Millie's earliest postdocs and her nominator for the award, in a 2007 *Boston Globe* article.[37] "She was very pleased to be a role model internationally," adds Laura Doughty.[38]

One of Millie's grandchildren, materials scientist Leora Dresselhaus-Marais, realized how well known her grandmother was only when she traveled to Paris with her high school classmates and discovered an enormous poster of Millie in the Charles de Gaulle airport following her receipt of the L'Oréal-UNESCO award. Growing up, Dresselhaus-Marais had been mostly shielded from Millie's fame. But after seeing the poster at the airport, "That was when I realized, okay, something's up here," she says.[39]

The Vannevar Bush Award, given by the National Science Board for exemplary public service in advancing science and technology, was also particularly meaningful for Millie, who won it in 2009. Bush, a longtime MIT professor of electrical engineering, was one of the nation's leading voices for publicly funded science and technology following World War II and was the lead driver behind the movement to establish the National Science Foundation in 1950 as a vehicle to support basic scientific research. At MIT, Bush also served as a vice president and dean of the engineering school, as well as chair and life member of the MIT Corporation, the Institute's governing body. MIT Building 13, Millie's longtime academic home, is named after Bush and to this day greets entrants with a large portrait of him in its lobby. His spirit even infiltrated Millie's workspace: upon joining the MIT faculty, Millie was assigned to a former Bush office, which brought her a great deal of inspiration.[40]

Millie was honored twice more with prestigious awards whose namesakes played significant roles in her life. The first was the Department of Energy's Enrico Fermi Award, named for her early mentor and marked with a special ceremony and meeting with President Obama at the White House. This was followed by the Materials Research Society's Von Hippel Award, named for another mentor and given for her materials science research, her leadership in energy and science policy, and her mentoring of young scientists.[41]

Additional honors during the 2010s—including election to the National Inventors Hall of Fame for her work on graphite intercalation and thermoelectricity, and receipt of the prestigious Institute of Electrical and Electronics Engineers (IEEE) Medal of Honor—only helped to solidify her place as a highly influential contributor to physics, materials science, and electrical engineering. And while she never ultimately took home a Nobel, another Scandinavian prize provided all the pomp, circumstance, and regalement she could ever have hoped for in celebration of her improbable career.[42]

First awarded in 2008 by the Kavli Foundation, a philanthropic organization established by Norwegian American entrepreneur Fred Kavli to advance and promote the scientific enterprise, the Kavli Prizes honor groundbreaking work in three distinct fields: astrophysics, nanoscience, and neuroscience. Given every other year for, respectively, the "biggest, smallest, and most complex" scientific advances, the prizes have quickly approached the cachet of their Nobel cousins. Winners are feted like celebrities and enjoy face time with at least one member of the Norwegian royal family. At an elaborate ceremony filled with black ties and ballroom gowns, they're handed a scroll, a gold medal, and a check for $1 million.[43]

Winning a Kavli Prize is a career-defining achievement for any scientist, and in Millie's case, the award was even more significant because she was the first solo recipient in any category. The eighty-one-year-old physicist was honored in part for developing "the foundation for our understanding of the influence of reduced dimensionality on the fundamental thermal and electrical properties of materials."[44] Says granddaughter Clara Dresselhaus, "There's something dumbfounding about coming home from school one day and being told, 'The king of Norway is presenting an award to your Grandma Millie.'"[45]

At the resplendent Kavli Prize ceremony in Oslo later that year, Millie dazzled in a cerulean blue dress that enveloped her in a dozen or so layers—a sartorial carbon fiber. She impressed many in the audience with her acceptance speech, during which she expressed deep appreciation for the award and announced she was donating her prize money to up-and-coming scientists.[46] "I was immediately taken with this brilliant person who had such a spirit of generosity," wrote Arlene Alda, who saw Millie for the first time that night and later interviewed her for a book, *Just Kids from the Bronx*.[47] For Millie, it was a night and an award to remember—one that both affirmed her life's work and inspired her to continue in new directions.

THE NEXT GENERATION

Throughout her final years, Millie happily participated in many research projects. Her last papers often related to two-dimensional materials, including graphene and phosphorene, a monolayer of phosphorus atoms analogous to graphene.[48] "She didn't slow down a lot," says Read Schusky, who managed Millie and Gene's professional affairs in their final years at MIT.[49] She did streamline her travel schedule quite a bit, however, and in what was perhaps the biggest sign of her age beginning to show, Millie eventually began arriving at MIT at 7:00 a.m. instead of 6:00.[50]

One particular delight in her later years was connecting with granddaughter Leora Dresselhaus-Marais, who completed a PhD program in physical chemistry at MIT. At the end of her undergraduate experience, Dresselhaus-Marais wasn't sure of her academic direction or where she'd go to graduate school. But when she visited Cambridge in spring 2012, Millie and Gene gave their granddaughter a peek of what life could be like if she chose MIT. "I started to realize that until I saw Millie in the science world, I would never really understand her," Dresselhaus-Marais says of her decision. "Getting to know Millie through MIT gave me a view [that] I would never have gotten otherwise."[51]

Together in Cambridge, the two planned weekly lunches. Their conversations focused largely on experiments, collaborations, and, indirectly, what it takes for a woman to navigate a career in science.[52] Historically, women had made great strides in the fifty-plus years since Millie began her career as an independent researcher. But even as laws and norms improved to protect them in the workplace, women in science and engineering have continued to report inequities; unequal promotions, exclusion in professional settings, difficulties surrounding family care, frequent unconscious biases, gender-based microaggressions, and sometimes flat-out harassment.

While Millie had enabled great strides for women in some respects, she admitted to underestimating the toll of persistent, and often invisible, inequities—such as demeaning comments or disparities in opportunities, expectations, pay, or lab space. She had been surprised at the outcome of a landmark 1999 report, initiated by MIT biology professor Nancy Hopkins and a number of colleagues, showing how female faculty in the MIT School of Science were systematically, albeit unintentionally, discriminated against—a report that had ripple effects throughout academia and prompted significant changes at MIT and elsewhere.[53]

"What Nancy was doing, and rightly so, is saying that we should have equality," Millie told the BBC in 2012. "I was just looking for tolerance, that we could have the opportunity to practice our field and work at an equal status once we were working. But what she was looking for was that we would have all the amenities that the men asked for and got."[54]

Millie also admitted to being somewhat hardened to behaviors falling under the umbrella of gender harassment: common microaggressions and microinsults, such as insinuations that women can't hack it in their field, or the tendency of men to be lauded for an idea that's ignored when first raised by a woman at the same meeting. "I think most women have experienced that, so that's an annoyance that you learn to live with," she told the BBC. "But the younger generation doesn't expect that and they don't like that."[55]

Even if some of their concerns weren't fully apparent, Millie continued to engage young scientists as a way to provide support. She encouraged young women in the field—including several of her own grandchildren with STEM aspirations—whenever possible. Late in her career, she took part in a series of Rising Stars workshops that assisted early-career women in science and engineering through networking and frank discussions on strategies to address persistent challenges.[56]

In her final years, Millie was presented with a unique opportunity to encourage women and girls to consider STEM careers. As part of a campaign to recruit and hire more women to work as scientists and engineers, General Electric invited Millie to star in a sixty-second commercial that asked: What if notable women in science were treated as celebrities with the same cachet as professional athletes, pop stars, and Hollywood actors? How might things change for women in the field? And how might industries and disciplines improve as a result? For the spot, which provided the opening scenes of this book, Millie played a world-famous science star: kids would dress like her, paparazzi and everyday pedestrians would angle for a photo wherever she went, and social media followers would show up for science lectures en masse. The video ended with GE committing to employ twenty thousand women within its science, engineering, and technology workforce by the year 2020.[57]

"They sent her the script, and she said, 'I don't really understand this,'" granddaughter Leora Dresselhaus-Marais explains. Millie couldn't see how acting like a celebrity was going to create a positive message about women in science, and she had only the foggiest idea what social media was or how it worked. "I said, 'Millie, you can't say no. . . . This is the type of thing that you need to do to reach this generation.'"[58]

Millie listened and went to film the ad. She felt awkward doing multiple takes, walking across a stage, and posing for selfies with actors. But the video had its intended effect: when it aired in 2017, it reached more than 30 million people on television and garnered a wildly positive response on social media.[59]

"The @generalelectric commercial with Millie Dresselhaus gives me one million goosebumps. Beautiful work," tweeted author Annie F. Downs. "Can you imagine what it would mean for girls if women like this were the celebrities to emulate?" tweeted Cindy Eckert, CEO of the Pink Ceiling. "That Millie Dresselhaus commercial gets me every time," tweeted former Michigan governor Jennifer Granholm, who later became the United States secretary of energy in the Joe Biden administration.[60]

"We wanted to shine a light on women in STEM and tell the story in a way that was going to engage people," Linda Boff, vice president and chief marketing officer at GE, told the Materials Research Society in 2017. "We were looking for someone with scientific accolades, but also someone who was likable, had charm, and personality, and who had a true commitment to women in the sciences. And Millie Dresselhaus checked those boxes so beautifully."[61]

Of course, the ad wasn't offering to address concerns about how women continue to face unique hurdles once they decide to enter science, engineering, or information technology, and it did not address the related problem of underrepresentation of groups including Black, American Indian/Indigenous, and Latinx professionals in STEM. Yet the project was in line with the ways in which Millie had, over the course of her career, tried to use her influence to inspire women to take their rightful place in these fields and then help to usher in positive change from within once there.

As of late 2020, the company had not released an update related to the hiring goal stated in Millie's ad. That year, however, GE named its first global chief diversity officer, as well as chief diversity officers for each of its nine business units — five of whom were women. A GE diversity report released in early

2021 noted that 22 percent of the company's global employees and 26 percent of its top leaders were women as of December 2020. "It is clear we have work to do," the report stated.[62] Julie Grzeda, director of global leadership programs and university relations for GE, told me via written statement that "while our portfolio and workforce have shifted over the past few years, our underlying commitment to inclusion and diversity has not changed. . . . Thanks in part to outreach campaigns like the video featuring Mildred 'Millie' Dresselhaus, we've engaged more women in tech with our university recruiting efforts, improved global hiring results, and are similarly proud to have a high percentage of women in our internships and leadership programs."[63]

Millie got to watch the final commercial well before its premiere on social media and its first showing on national TV in February 2017, when it aired during a telecast of *Saturday Night Live*. GE also planned to air it during the Eighty-Ninth Academy Awards, and the company invited Millie to attend the Oscars that night, a star of science mingling on the red carpet with the biggest stars of the silver screen. But Millie never made it to the Oscars, and she never got to see the impact that the video ultimately had or to understand how and why this last public act would become her most widely known.

LOSING A LEGEND

Just as MIT students returned from their winter break in early February 2017, the Boston area braced for a blizzard that threatened to bring all business and scholarly activities to a screeching halt. Of course, the forecast didn't rattle Millie, who had planned her week as usual and arrived at her lab raring to go.

It turned out to be her final arrival at MIT, her beloved home away from home.

Later that morning, Millie felt very unwell. She was eventually admitted at Mount Auburn Hospital, on the western edge of Cambridge. According to doctors there, she had suffered a stroke and needed to remain on bed rest until further notice.[64]

A couple of days later, one of her former students, Marcie Black, began to worry when Millie failed to return an email. "After a few hours, I already knew something was wrong," Black later noted. "She always returned emails within an hour."[65]

Family and close friends came to visit Millie over the next two weeks. On Monday, February 20, the world lost a legend as Mildred Dresselhaus died peacefully, surrounded by her loved ones. She was laid to rest at nearby Mount Auburn Cemetery, less than two hundred yards from the grave of Buckminster Fuller, namesake of the very molecules she had helped to discover and knew so intimately. A marker featuring carbon hexagons now reads: "Cherished Wife, Mother, Grandmother; Physicist & MIT Professor; Queen of Carbon."[66] (Gene Dresselhaus joined Millie four and a half years later; he died at 91 on September 29, 2021.)

The following weekend, Millie's GE ad aired on prime-time television, as planned, during the Academy Awards broadcast. For those like me who not only knew about her legacy but about her passing just days before, seeing the minute-long spot in the midst of the evening's jubilation spurred a flood of emotions. "I cried a river in that moment," says MIT professor Sangeeta Bhatia, a colleague who had long considered Millie a shining light. "It was such a good celebration of her."[67]

In Millie's stead, two of her granddaughters with career ambitions in the STEM fields, Leora Dresselhaus-Marais and Clara Dresselhaus, walked the red carpet as guests of GE. "I thought that brought it full circle," Dresselhaus-Marais says. "Let's use this as a demonstration of the next generation of female scientists."[68]

Following her death, tributes poured in from around the globe in honor of the late Mildred Dresselhaus.

"She was an icon for many virtues—science, women, truth, and service, among others," physicist George Crabtree wrote in the *Materials Research Society Bulletin*. "She represented these things so well and so naturally that many of us took them for granted."[69]

"Millie was such an inspiration for so many women students, staff and faculty at MIT, and such a career support system for me," said former student Aviva Brecher for an oral history project later that year. "She got her students and mentees so many good jobs in industry, and created such an awareness of the importance of science policy that MIT started an office in Washington."[70]

"Brilliant mind, deep insight, perfect scientific intuition—these are just a few of her characteristics as a scientist," Elena Rogacheva, professor of physics at the National Technical University Kharkiv Polytechnic Institute in the Ukraine and a former collaborator, wrote in a *Physics Today* tribute. "Millie's work was related to energy problems, but she had an inexhaustible source of energy inside. It seemed it would last forever."[71]

"The most important thing I learned from Millie was not about materials science, group theory, Raman spectroscopy, or nanotechnology. It was about always looking at the positive, constructive side of things," said former student Ado Jorio in *Physics Today*. "Millie would never point [out] or criticize the negative aspect of a paper—she would always point to its strength. Millie would hear people's complaints about the structure of the building, about politics, about whatever, and reply with a comment about the good side of it. When there was no way to see a good side, she would just say 'better times will come.' Life with Millie was always very, very, very hard working, but with pleasure, with happiness, with a smile."[72]

"I hope, Millie-sensei, you can perhaps now write papers in heaven," said Japanese physicist and collaborator Morinobu Endo, at a tribute event in late 2017. "Please watch over

Gene-sensei and your family . . . and also world scientists for further development of their studies [in carbon and other materials], as you are the common mother of their research."[73]

AN IMPROBABLE LIFE

A year and a half after her passing, MIT opened its doors to a gleaming new nanoscience and nanotechnology research facility at the heart of its campus. Dubbed MIT.nano, the $400 million project was a long time in coming. While Millie missed out on the grand opening, she was very much looking forward to its completion and to the start of a new generation of nanoscale endeavors at the Institute that would seek to expand humanity's understanding of physics, chemistry, materials science, energy, biology, and more.[74]

According to granddaughter Shoshi Dresselhaus-Cooper, Millie also looked at MIT.nano as an extension of her legacy at the Institute. In late 2019 the courtyard between MIT's Infinite Corridor and the southern facade of MIT.nano was dedicated in her memory. Dubbed the Improbability Walk, the space makes a nod to Millie's unlikely rise to international prominence from her humble beginnings in New York City and encourages those who might serve as mentors to take time to get to know younger colleagues and students whose life path might be immeasurably enhanced by an encouraging word.[75]

In just the few years since her passing, new advances from her colleagues already bear the signature of Millie's work and are branching into ever-more fascinating directions. Graphene, for example, remains one of the hottest topics in science. In the early and mid-2010s, Millie worked on what she and others called "misoriented graphene"—twisting sheets of graphene such that their honeycomb patterns are slightly misaligned when superimposed.[76] Doing so, Millie and others predicted, could introduce "interesting patterns" that might lead to useful properties.[77] In 2018, Millie's MIT colleague Pablo Jarillo-Herrero realized this idea when he and others discovered that if two graphene sheets are combined into a superlattice, aligned at a "magic angle" of 1.1 degrees, the system can become either superconducting or insulating. The development was hailed as a major discovery, marking a jumping-off point for a subfield now known as "twistronics," and was named the *Physics World* Breakthrough of the Year.[78]

For another young scientist, Millie's work lives on more viscerally, as the physical backdrop to a new career. Canadian-born nanotechnologist Farnaz Niroui never imagined she'd be following, quite literally, in the footsteps of one of the most renowned researchers of nanoscale materials. But the MIT assistant professor, a recent addition to the MIT Department of Electrical Engineering and Computer Science faculty, takes daily inspiration from her workspace—Millie's office at MIT. Niroui's research focuses on making matter at the atomic level easier to manipulate and process so that scientists and engineers can investigate and exploit its chemical and physical properties. While a PhD student at MIT, the highly talented Niroui garnered multiple offers to join the faculties of other prestigious universities around North America. But MIT could give her something no one else could: the opportunity to sit every day where Millie sat and to make a career in shaping the nanoscience and nanotechnology facility that the Queen of Carbon had been so excited about. "I feel like I can be a little part of a legacy," Niroui says.[79]

Niroui's work doesn't only overlap in her nanoscale research; she's also following Millie's example in supporting women in science and engineering and in making nanoscience and nanotechnology more accessible to everyone. Among other things, she has organized workshops for women in STEM who are planning careers as tenure-track professors, and she led the development of a Perspectives in Nanotechnology speaker series at MIT, as well as a new Mildred S. Dresselhaus Lecture Series at MIT. "I knew Millie as a grad student since our offices were very close to each other, and I always looked up to her," Niroui told the MIT Alumni Association in 2018. "Being able to be in her office as I start my own independent academic career is extremely inspiring and very valuable to me."[80]

Millie's immeasurable contributions will continue to have an impact on nanoscience and nanotechnology—in both the research realm and the people she mentored, trained, and encouraged in other ways. Her investigations into the properties of semimetals were second-to-none and influenced countless lines of inquiry into nanomaterials. She was a pivotal figure in reigniting the field of thermoelectrics as a way to improve energy conversion. She was highly influential as an educator: as an author of important textbooks and review articles, as one who often summarized research conferences and information sessions, and as a student adviser to hundreds and a classroom teacher to thousands. She was a valued leader in the administration of science within organizations and in the US government. And she remained steadfast in working to provide improved opportunities for women and individuals from other underrepresented groups in science and engineering.

In the end, Millie's remarkable history serves as a reminder of how a single person can overcome great odds while also striving to better her world in ways big and small. A lot had to go right for Millie to end up where she did. But it was her own persistence, coupled with her brilliance and her immense care for others, that in fact turned Millie into one of the preeminent scientists of her time.

It's clear that Millie was an explorer at heart. From her earliest days in New York City, she sought out new frontiers in both personal experiences and answers to life's ceaseless questions. She thrived in a career where she could contribute to basic research, adding to our understanding of carbon and other materials. And she derived much joy from being able to let her mind wander and follow any lead that piqued her curiosity.[81] "It's a great privilege to be able to participate in this great explosion of knowledge—even more of a privilege to teach it," Millie once stated.[82] "I think that scientists have a goal of discovering the unknown, and that's a forever-goal. It keeps you alert and excited about each day," she added in 2013.[83]

Millie also, of course, felt deeply about giving back—to students, to her research community, to society at large. "When I started out my career, I thought I had no career. And the fact that I have had a career is because of society. So, I feel that I should pay back for that," she noted in 2007.[84]

Mildred Dresselhaus more than made up for any gifts she received throughout her life. She was, in her eighty-six-plus years, a giver through and through—of her time, her intellect, her energy, her love, and her enthusiasm.

In one of her final interviews, the Queen of Carbon made sure that her words would provide a wake on which others might surf future waves of discovery. "We need new science and we need new ideas, and there's plenty of room for young people to come in and have careers discovering those new ideas," she declared. "Life is very interesting in this lane. Come and join me!"[85]

ACKNOWLEDGMENTS

I was a young science reporter at *Discover* magazine in summer 2002 when I received one of my most memorable research assignments: a feature article highlighting fifty individuals the editors called the "most important women in science" at that time. Highly interested in the history of women in science, I gleefully studied each person on the list and contacted her to be sure our information was accurate. Sandwiched between particle physicist Persis Drell and oceanographer Sylvia Earle in the final layout, the inimitable Mildred "Millie" Dresselhaus beamed her usual inviting grin.

When I arrived in 2014 as an editor in the MIT News Office, I began to understand Millie's influence more intimately. It didn't take long to hear her name mentioned with reverence in the Institute's hallowed halls, and I was awed by the research she continued well into her eighties. I had a chance to meet the esteemed Institute Professor Emerita later that year. On an overcast November afternoon, just days before she picked up her Presidential Medal of Freedom from Barack Obama, I trekked to her office for what I expected would be a quick handoff of the LEGO minifigure I'd created in her likeness. Instead, she warmly hosted me for at least half an hour, during which we chatted about worldly travels, women in science, and growing up in New York, before posing for a photo and placing the minifigure among her prized awards.

Yet it wasn't until she passed away in 2017 at age eighty-six that I started to discover the real Millie Dresselhaus. Piecing together her life from decades of interviews and profiles, as well as new interviews with family, colleagues, students, and others, the scientist, mentor, administrator, mother, and grandmother I came to know was a prismatic rainbow of colors in comparison to the single bright wavelength I first glimpsed back in 2002.

For inviting me to embark on an epic journey through Millie's life, I am first and foremost indebted to my MIT Press colleagues—in particular the magnificent Jermey Matthews, who walked with me over the bridge to first authorship, providing key guidance and support along the way; Amy Brand for encouraging the project through its various stages; Haley Biermann for procedural assistance through to production; four anonymous reviewers, whose helpful comments made the manuscript stronger; and Gita Manaktala, Judith Feldmann, Beverly Miller, Susan Clark, Margarita Encomienda, Sean Reilly, Kate Elwell, Jay Martsi, Nicholas DiSabatino, Jess Pellien, Bill Smith and his sales team, and all other editors, designers, and marketing managers who helped launch the book out into the world. I also owe a great deal to donors who backed the project through the MIT Press Fund for Diverse Voices, which supports the writing of self-identified women, nonbinary individuals, and other underrepresented groups in the STEM fields.

The extended Dresselhaus family provided historical information, eye-opening photos, and poignant tidbits about Millie and Gene. I especially appreciate the contributions of Shoshi Dresselhaus-Cooper, who shared insights, archival documents, photo information, and other resources, including a highly detailed time line of Millie's career. Many thanks are also due to Marianne Dresselhaus Cooper for inviting me into her family's world and to Gene, Paul, Eliot, Carl, Elizabeth, and Clara Dresselhaus as well as Leora Dresselhaus-Marais and Geoffrey Cooper for engaging with me, and for sharing stories and experiences.

Laura Doughty was the Dresselhaus's administrative assistant for two decades, and in that time served in many ways as a sustained wind beneath Millie and Gene's wings. Thank you, Laura, for your reminiscences, and for imparting your view of Millie's magic.

Many others shared with me their Millie memories. I thank Read Schusky, Clarence Williams, Gang Chen, Jing Kong, Aviva Brecher, Elizabeth Stewart, Barbara Jacobs, Marcie Black, Farnaz Niroui, Mario Hofmann, Sangeeta Bhatia, Cherry Murray, Cyrus Mody, Sherwin Lehrer, and Laura Roth for their remembrances.

A number of individuals provided early encouragement and valuable feedback. I very much appreciate Emily Neill, Joseph Martin, Traci Swartz, Arthur Eisenkraft, Tom Gearty, Seth Mnookin, Alice Dragoon, Deborah Blum, Peter Dunn, Deborah Chung, Marion Reine, Sarah Simon, Coleen Smith, Lauren Gravitz, and especially Jennifer Chu, Daniel Hudon, and Emily Hiestand for their considerable guidance.

I give heartfelt thanks to Carolyne Van Der Meer for allowing me to publish her moving ode to Millie and to Gillian Dreher, Theresa Machemer, and Ele Willoughby for lending their beautiful Millie portraits.

I am grateful to my MIT News Office colleagues Steve Bradt, Kimberly Allen, and especially Kathy Wren, who provided encouragement and filled in for me on the many vacation days I spent working on this project. Thank you also to the MIT Libraries staff, particularly Myles Crowley, Nora Murphy, and Elizabeth Andrews in MIT Distinctive Collections, and to staffers at the Schlesinger Library at the Harvard Radcliffe Institute and the Hunter College High School library. I also appreciate Doreen Morris in the MIT Office of the Provost, who provided helpful archival information about the Abby Rockefeller Mauzé professorship; Lydia Snover in MIT Institute Research for data on women at MIT; Rachel Kemper, Anne Stuart, Irene Yong Rong Huang, and Stephen Salk for MIT department-level and human resources guidance; and Robert M. Gray and Debbie Douglas for insights relating to the history of women at MIT. I also thank my MIT colleagues who assisted in making this project logistically possible for me, as well as collaborators at Lumina Datamatics who supplied important work on photos, graphics, and illustrations, and David Luljak for work on the index.

I am indebted to numerous individuals who assisted in the care of my family while I worked on this book and on my other professional endeavors. I share my gratitude with each of you, as you make everything possible.

I could not have written this book without the extraordinary support of my family, especially my parents, Rosa and Lou, who lent encouragement and regular stretches during which I could dig in and write; my brother, Jordi, who gave me the fortitude to take on such a significant project; the Weinstars, for daily inspiration; and my beloved Minou, a regular source of purrs during countless late nights of writing. Gràcies per tot.

To my little one: Millie's story is your story as well, for it grew along with you in your first few years. Thank you for being the light of my life.

TIME LINE OF KEY MILESTONES

November 11, 1930—Mildred Spiewak is born in Brooklyn, New York

1948—Graduates Hunter College High School, New York, New York

1951—Graduates Hunter College (BA)

1951–1952—Fulbright Fellow, Cambridge University

1953—Graduates Radcliffe College (MA)

1958—Marries Gene Dresselhaus

1958—Graduates University of Chicago (PhD)

1958–1960—National Science Foundation Postdoctoral Fellow, Cornell University

1959—Gives birth to daughter Marianne

1960—Begins work as a research scientist at MIT Lincoln Laboratory

1961—Gives birth to son Carl

1963—Gives birth to son Paul

1964—Gives birth to son Eliot

1967–1968—Abby Rockefeller Mauzé Visiting Professor, MIT Department of Electrical Engineering

1968—Appointment as tenured professor in the MIT Department of Electrical Engineering

1974–1972—Associate head of the MIT Department of Electrical Engineering

1974—Election to National Academy of Engineering

1977–1983—Director of MIT Center for Materials Science and Engineering

1983—Joint appointment to faculty of MIT Department of Physics

1984—President of the American Physical Society

1985—Appointment as MIT Institute Professor

1985—Election to National Academy of Sciences

1986—MIT James R. Killian Jr. Faculty Achievement Award

1990—US National Medal of Science

1997–1998—President of the American Association for the Advancement of Science

2000–2001—Director of the US Department of Energy Office of Science

2003–2008—Chair of the American Institute of Physics Governing Board

2005—Heinz Family Foundation Heinz Award

2007—L'Oreal-UNESCO Award for Women in Science

2008—American Association of Physics Teachers Oersted Medal

2009—National Science Board Vannevar Bush Award

2009—Elected fellow of the Materials Research Society

2012—US Department of Energy Enrico Fermi Award

2012—Kavli Prize

2013—Materials Research Society Von Hippel Award

2014—Inducted into the National Inventors Hall of Fame

2014—US Presidential Medal of Freedom

2015—IEEE Medal of Honor

February 20, 2017—Mildred S. Dresselhaus dies in Cambridge, Massachusetts

NOTES

PROLOGUE

1. Aditi Risbud, "Millie Dresselhaus: Our Science Celebrity," *MRS Bulletin* 42, no. 11 (2017): 788, https://doi.org/10.1557/mrs.2017.262.

2. "Carbon," Periodic Table of Elements, Los Alamos National Laboratory, https://periodic.lanl.gov/6.shtml.

3. Nick Bilton, "Bend It, Charge It, Dunk It: Graphene, the Material of Tomorrow," *New York Times*, April 13, 2014, https://bits.blogs.nytimes.com/2014/04/13/bend-it-charge-it-dunk-it-graphene-the-material-of-tomorrow.

4. "1954: Silicon Transistors Offer Superior Operating Characteristics," Computer History Museum, Mountain View, CA, https://www.computerhistory.org/siliconengine/silicon-transistors-offer-superior-operating-characteristics.

5. Andrew Grant, "Mildred Dresselhaus (1930–2017)," *Physics Today*, February 23, 2017, https://physicstoday.scitation.org/do/10.1063/PT.5.9088/full/.

CHAPTER 1

1. Eamon Loingsigh, "Sands Street Station," *ArtofNeed*, June 9, 2013, https://artofneed.wordpress.com/2013/06/09/break-for-edit; "History

of the Yard," Brooklyn Navy Yard, https://brooklynnavyyard.org/about/history.

2. Andrew J. Sparberg, *From a Nickel to a Token: The Journey from Board of Transportation to MTA* (New York: Fordham University Press, 2015), 45–47; "Pedestrian Counts," Downtown Alliance, https://www.downtownny.com/pedestrian-counts; "Cyclist Counts on East River Bridge Locations," New York City Department of Transportation, https://www1.nyc.gov/html/dot/downloads/pdf/east-river-bridge-24hr-cyclist-count-oct2019.pdf; "2016 New York City Bridge Traffic Volumes," New York City Department of Transportation, February 2018, http://www.nyc.gov/html/dot/downloads/pdf/nyc-bridge-traffic-report-2016.pdf.

3. Marianne Dresselhaus Cooper, interview by author, Arlington, MA, April 27, 2018; Mildred Dresselhaus, interview by Kelsey Irvin, 2013, transcript, Oral History Program, IEEE History Center, Hoboken, NJ, https://ethw.org/Oral-History:Mildred_Dresselhaus; US Census Bureau, "1930 Federal Population Census," prepared by Ancestry.com; Marianne Dresselhaus Cooper, emails to author, May 31, June 8, 2020.

4. Mark Anderson, "The Queen of Carbon," *IEEE Spectrum* 52, no. 5 (2015): 50–54; Mildred Dresselhaus, interview, 2004, by Magdolna Hargittai and István Hargitti, *Candid Science IV: Conversations with Famous Physicists* (London: Imperial College Press, 2004), 546–569; Marianne Dresselhaus Cooper, interview, 2018; Mildred Dresselhaus, interview by Shirlee Sherkow, 1976, transcript, Project on Women as Scientists and Engineers, MIT Libraries Distinctive Collections, Cambridge, MA, 2–3.

5. Mildred Dresselhaus, interview, 2004, 546–569; Marianne Dresselhaus Cooper, email to author, June 8, 2020.

6. "Irving Spiewak," Find a Grave, March 3, 2009, https://www.findagrave.com/memorial/34387569/irving-spiewak; Mildred Dresselhaus, interview, 2013.

7. Robert J. Schoenberg, *Mr. Capone: The Real—and Complete—Story of Al Capone* (New York: HarperCollins, 2001), 19.

8. Marianne Dresselhaus Cooper, interview, 2018; Loingsigh, "Sands Street Station."

9. History.com Editors, "Great Depression History," History.com, October 29, 2009, https://www.history.com/topics/great-depression/great -depression-history; Marianne Dresselhaus Cooper, interview, 2018; Mildred Dresselhaus, interview, 1976, 10–11.

10. Mildred Dresselhaus, interview, 1976, 6–8, 92–93; Mildred Dresselhaus, "Memories from a Life in Physics," *MIT Physics Annual* (2009): 46–51.

11. Mildred Dresselhaus, interview, 1976, 6–7.

12. Mildred Dresselhaus, interview by Arlene Alda, *Just Kids from the Bronx: Telling It the Way It Was: An Oral History* (New York: Holt, 2015), 42–45; Mildred Dresselhaus, interview, 1976, 6–7.

13. Alda, *Just Kids from the Bronx*, 42–45; Mildred Dresselhaus, interview, 1976, 6–7.

14. Mildred Dresselhaus, interview by the US Department of Energy Office of Science, "Fermi Award Winners: Q&A," US Department of Energy Office of Science, June 6, 2012, https://web.archive.org/web /20150908034322/https://science.energy.gov/news/featured-articles /2012/06-06-12/; "History of the Bronx," Yes the Bronx, http:// yesthebronx.org/about/history-of-the-bronx/, Mildred Dresselhaus, interview, 1976, 7.

15. "Industrial Depression—Unemployment—Destitution! Idleness and Want Drive Men to Crime and Suicide! Desperate Situation Demands Serious Consideration!" *Minnesota Union Advocate*, March 5, 1931, 1.

16. Mildred Dresselhaus, interview, 1976, 7.

17. Mildred Dresselhaus, interview , 2015, 42–45; Mildred Dresselhaus, interview, 1976, 7.

18. Mildred Dresselhaus, "Memories," 46–47; Mildred Dresselhaus, interview by Jenni Murray, "The Age of Reason," BBC, December 2012, https://www.bbc.co.uk/sounds/play/p012bp6b; Mildred Dresselhaus, interview, 1976, 92; Mildred Dresselhaus, interview by Paul S. Weiss, "A Conversation with Prof. Mildred Dresselhaus: A Career in Carbon Nanomaterials," *ACS Nano* 3, no. 9 (2009): 2438.

19. Mildred Dresselhaus, interview, 1976, 91.

20. Mildred Dressclhaus, interview, 2015; Mildred Dresselhaus, "Memories"; Shoshi Dresselhaus-Cooper, email to author, April 9, 2018; Google Maps with Street View; "FDNY Rescue Company 3," *Architect*, July 17, 2012, https://www.architectmagazine.com/project-gallery/fdny-rescue-company-3-320; "Perrigo Co PLC," Bloomberg, https://www.bloomberg.com/profile/company/PRGO:US; "Perrigo New York, Inc." Vault, https://www.vault.com/company-profiles/pharmaceuticals-and-biotechnology/perrigo-new-york-inc.

21. "NYCityMap," City of New York, http://maps.nyc.gov/doitt/nycitymap/; Mildred Dresselhaus, interview, June 6, 2012; "History of the Bronx," Yes the Bronx; "Prohibition," History.com, Oct. 29, 2009, https://www.history.com/topics/roaring-twenties/prohibition.

22. Mildred Dresselhaus, interview, 1976, 106.

23. Mildred Dresselhaus, interview, 1976, 108.

24. Mildred Dresselhaus, interview, June 2012; Anderson, "Queen of Carbon"; Mildred Dresselhaus, "Mildred Dresselhaus Biography," Kavli Prize, http://kavliprize.org/sites/default/files/%25nid%25/auto biagraphies_attachments/Mildred_Dresselhaus_Biography_0.pdf; Mildred Dresselhaus, interview, 2015.

25. Mildred Dresselhaus, interview, 2015.

26. Mildred Dresselhaus, interview, 1976, 106–107; Mildred Dresselhaus, interview, December 2012.

27. Mildred Dresselhaus, interview, December 2012.

28. Mildred Dresselhaus, interview, 1976, 3–4, 10–11; Kimberly Amadeo, "Unemployment Rate by Year since 1929 Compared to Inflation and GDP," *The Balance*, https://www.thebalance.com/unemployment-rate-by-year-3305506; Mildred Dresselhaus, interview, 2015, 44; Leora Dresselhaus-Marais via Marianne Dresselhaus Cooper, email, June 8, 2020.

29. Mildred Dresselhaus, interview by Harry Kroto, Vega Science Trust, 2001, http://www.vega.org.uk/video/programme/20.

30. Mildred Dresselhaus, interview, 2001.

31. Mildred Dresselhaus, interview, 1976, 10; Mildred Dresselhaus, interview, 2001.

32. Mildred Dresselhaus, interview, 2001.

33. Mildred Dresselhaus, interview, 2015, 42–45.

34. Shoshi Dresselhaus-Cooper, emails to author, May 3, 2018; Mildred Dresselhaus, interview, 1976, 82.

35. Shoshi Dresselhaus-Cooper, email, May 3, 2018.

36. Mildred Dresselhaus, interview, 1976, 10.

37. Shoshi Dresselhaus-Cooper, email to author, April 9, 2018; Mildred Dresselhaus, interview, 1976, 11.

38. Mildred Dresselhaus, interview, 1976, 11.

39. "Modern Times," Rotten Tomatoes, https://www.rottentomatoes.com/m/modern_times; Marianne Dresselhaus Cooper, email, June 8, 2020.

40. Shoshi Dresselhaus-Cooper, email to author, April 9, 2018; Marianne Dresselhaus Cooper, email, June 8, 2020.

41. Mildred Dresselhaus, interview, 2015, 42–45.

42. Mildred Dresselhaus, interview, June 2012.

43. "The History of Greenwich House," Greenwich House, https://www.greenwichhouse.org/history; Irving Spiegel, "School's Concert a Story of Music," *New York Times*, March 4, 1950, 19.

44. Mildred Dresselhaus, interview, June 2012.

45. Keith O'Brien, "Pioneering Woman Physicist, Cited for Her Research, Mentoring," *Boston Globe*, March 5, 2007, 19; Mildred Dresselhaus, interview, 1976, 7, 92.

46. Mildred Dresselhaus, interview, 2015, 42–45; Mildred Dresselhaus, interview, 1976, 15.

47. Shoshi Dresselhaus-Cooper, email to author, May 15, 2018; Mildred Dresselhaus, interview, 1976, 15.

48. Mildred Dresselhaus, interview, 2015, 43.

49. "Fantasia in Eight Parts: 'The Sorcerer's Apprentice,'" Walt Disney Family Museum, Aug. 2, 2012, https://www.waltdisney.org/blog/fantasia-eight-parts-sorcerers-apprentice; "Fantasia in Eight Parts: 'Toccata and Fugue in D minor,'" Walt Disney Family Museum, Aug. 28, 2012, https://www.waltdisney.org/blog/fantasia-eight-parts-sorcerers-apprentice.

50. Marianne Dresselhaus Cooper, email, June 8, 2020; Shoshi Dresselhaus-Cooper, email to author, April 9, 2018; Mildred Dresselhaus, interview, 2015, 42–45.

51. Marianne Dresselhaus Cooper, interview, 2018; Greenwich House, "The History of Greenwich House."

52. Eleanor Roosevelt, "My Day, July 26, 1939," Eleanor Roosevelt Papers Digital Edition (2017, https://www2.gwu.edu/~erpapers/myday/displaydoc.cfm?_y=1939&_f=md055328.

53. "They Shall Have Music," TCM.com, https://www.tcm.com/watchtcm/movies/92847/They-Shall-Have-Music.

54. Eleanor Roosevelt, "My Day, July 27, 1939," Eleanor Roosevelt Papers Digital Edition (2017), https://www2.gwu.edu/~erpapers/myday/displaydoc.cfm?_y=1939&_f=md055329.

55. David Dworkin, "The Legacy of Mary Kingsbury Simkhovitch," National Housing Conference, March 19, 2019, https://www.nhc.org/the-legacy-of-mary-kingsbury-simkhovitch/; Mary Kingsbury Simkhovitch Papers, 1852–1960; A-97, Schlesinger Library, Radcliffe Institute, Harvard University, Cambridge, MA; Shoshi Dresselhaus-Cooper, email to author, April 9, 2018.

56. Shoshi Dresselhaus-Cooper, email to author, May 16, 2018.

57. Mildred Dresselhaus, interview, 1976, 9.

58. Mildred Dresselhaus, interview, 2004, 548; Paul De Kruif, *Microbe Hunters* (New York: Harcourt, 1926).

59. Cherry Murray, email to author, July 22, 2018; Eve Curie, *Madame Curie: A Biography* (New York: Da Capo Press, 2001 reissue); Eugene Straus, *Rosalyn Yalow Nobel Laureate: Her Life and Work in Medicine* (New York: Plenum Press, 1998), 66.

60. Mildred Dresselhaus, interview, 2001.

61. Mildred Dresselhaus, interview, 2004.

62. Mildred Dresselhaus, interview, 2004.

63. Mildred Dresselhaus, interview, 1976, 107–108.

64. Mildred Dresselhaus, interview, 1976, 108–109.

65. Mildred Dresselhaus, interview, 1976, 108–109.

66. Mildred Dresselhaus, interview, June 2012; Mildred Dresselhaus, interview, 1976, 91–94; Mildred Dresselhaus, "Memories," 47; Mildred Dresselhaus, interview, September 2009, 2438–2439.

67. Mildred Dresselhaus, interview, 1976, 4–5, 48; "Irving Spiewak," Find a Grave; Rick Seltzer, "Free Again—in 10 Years," *Inside Higher Ed*, March 16, 2018, https://www.insidehighered.com/news/2018/03/16 /cooper-union-plans-restore-free-undergraduate-tuition-decade.

68. Mildred Dresselhaus, interview, 1976, 25–26; "About," Bronx High School of Science, https://www.bxscience.edu/apps/pages/index .jsp?uREC_ID=219378&type=d&termREC_ID=&pREC_ID=433038 &hideMenu=0; "Stuyvesant History—Enter the First Girls," Stuyvesant High School Alumni Association, https://www.stuyalumni.org/news /stuyvesant-history-enter-the-first-girls; Laurie Gwen Shapiro, "How a Thirteen-Year-Old Girl Smashed the Gender Divide in American High Schools," *New Yorker*, January 26, 2019, https://www.newyorker .com/culture/culture-desk/how-a-thirteen-year-old-girl-smashed-the -gender-divide-in-american-high-schools; "School History—History of Tech," Brooklyn Technical High School, https://www.bths.edu/school _history.jsp.

69. Mildred Dresselhaus, interview, September 2009, 2439.

70. Mildred Dresselhaus, interview, 1976, 26–27.

71. Mildred Dresselhaus, interview, 1976, 24.

72. Mildred Dresselhaus, interview, 1976, 27.

73. Mildred Dresselhaus, interview, 1976, 26.

74. Mildred Dresselhaus, interview, 2015, 42–45.

CHAPTER 2

1. Mildred Dresselhaus, interview by Shirlee Sherkow, 1976, transcript, Project on Women as Scientists and Engineers, MIT Libraries Distinctive Collections, Cambridge, MA, 101.

2. Mildred Dresselhaus, interview, 1976, 101.

3. Betty Stewart, email to author, June 19, 2018; Mildred Dresselhaus, interview, 1976, 100–101, 107.

4. Mildred Dresselhaus, interview by Arlene Alda, *Just Kids from the Bronx: Telling It the Way It Was: An Oral History* (New York: Holt, 2015), 42–45; Mildred Dresselhaus, interview, 1976, 16, 110.

5. Marianne Dresselhaus Cooper, interview by author, Arlington, MA, April 27, 2018; Kimberly Amadeo, "Unemployment Rate by Year since 1929 Compared to Inflation and GDP," *The Balance*, September 17, 2020, https://www.thebalance.com/unemployment-rate-by-year -3305506.

6. Mildred Dresselhaus, interview, 1976, 81–82.

7. Mildred Dresselhaus, interview, 2015, 42–45; Mildred Dresselhaus, interview, 1976, 82–83.

8. Mildred Dresselhaus, interview, 1976, 14, 107.

9. Mildred Dresselhaus, interview, 1976, 14.

10. Mildred Dresselhaus, interview, 1976, 18.

11. Mildred Dresselhaus, interview, 1976, 17.

12. Mildred Dresselhaus, interview, 1976, 17.

13. Mildred Dresselhaus, interview, 1976, 17.

14. Mildred Dresselhaus, interview, 1976, 17–18.

15. Mildred Dresselhaus, interview, 1976, 18.

16. Mildred Dresselhaus, interview, 1976, 15.

17. Mildred Dresselhaus, interview, 1976, 19.

18. Mildred Dresselhaus interview, 1976, 28; Mildred Dresselhaus interview in US Department of Energy Office of Science, "Fermi Award Winners: Q&A," US Department of Energy Office of Science, June 6, 2012, https://web.archive.org/web/20150908034322/https://science .energy.gov/news/featured-articles/2012/06-06-12/.

19. "Hunter College High School Address by Mildred Dresselhaus," Hunter College High School, December 1, 2009, https://www.youtube .com/watch?v=zTe6mAvWB2M.

20. Mildred Dresselhaus interview, 1976, 28.

21. Elizabeth Balletto Stewart, email to author, June 19, 2018.

22. Stewart, email.

23. Mildred Dresselhaus interview, 1976, 11–12.

24. "Value of $5 from 1946 to 2021, Inflation Calculator," Official Inflation Data, Alioth, https://www.officialdata.org/us/inflation/1946 ?amount=5.

25. Mildred Dresselhaus interview, 1976, 11.

26. Mildred Dresselhaus interview, 1976, 12.

27. Mildred Dresselhaus interview, 1976, 93; "Pertussis Cases by Year (1922–2018)," US Centers for Disease Control and Prevention, https:// www.cdc.gov/pertussis/surv-reporting/cases-by-year.html.

28. C. G. Shapiro-Shapin, "Pearl Kendrick, Grace Eldering, and the Pertussis Vaccine," *Emerging Infectious Diseases* 16, no 8 (2010): 1273–1278, https://dx.doi.org/10.3201/eid1608.100288; Jean-Marc Cavaillon, Sansonetti Philippe, and Goldman Michel, "100th Anniversary of

Jules Bordet's Nobel Prize: Tribute to a Founding Father of Immunology," *Frontiers in Immunology*, September 11, 2019, https://doi.org/10.3389/fimmu.2019.02114; Brian Shaw, "Leila Denmark (1898–2012)," New Georgia Encyclopedia, April 24, 2013, https://www.georgiaencyclopedia.org/articles/science-medicine/leila-denmark-1898-2012; Shift7, "#20for2020: Pearl Kendrick, Grace Eldering, and Loney Clinton Gordon Developed the Whooping Cough and Single Dose DTP Vaccines," Amy Poehler's Smart Girls, January 12, 2020, https://amysmartgirls.com/20for2020-pearl-kendrick-grace-eldering-and-loney-clinton-gordon-developed-the-pertussis-and-c035f2858d6.

29. Mildred Dresselhaus interview, 1976, 93.

30. Margaret Nash and Lisa Romero, "Citizenship for the College Girl: Challenges and Opportunities in Higher Education for Women in the United States in the 1930s," *Teachers College Record* 114, no. 2 (2012): 5–6, https://www.academia.edu/12116246/_Citizenship_for_the_College_Girl_Challenges_and_Opportunities_in_Higher_Education_for_Women_in_the_United_States_in_the_1930s; Scott A. Ginder, Janice E. Kelly-Reid, and Farrah B. Mann, "Postsecondary Institutions and Cost of Attendance in 2017–18; Degrees and Other Awards Conferred, 2016–17; and 12-Month Enrollment, 2016–17," US Department of Education National Center for Education Statistics, November 2018, https://nces.ed.gov/pubs2018/2018060REV.pdf, table 4.

31. Mildred Dresselhaus interview, 1976, 29; Rebecca Onion, "Unclaimed Treasures of Science," *Slate*, July 13, 2014, https://slate.com/technology/2014/07/women-in-science-technology-engineering-math-history-of-advocacy-from-1940-1980.html.

32. Mildred Dresselhaus interview in US Department of Energy Office of Science.

33. Mildred Dresselhaus interview, June 2012; Mildred Dresselhaus interview, 1976, 11–12.

34. Mildred Dresselhaus interview, 1976, 29–30.

35. Mildred Dresselhaus interview, 1976, 30–31.

36. Commencement Program, Hunter College High School, February 3, 1948.

37. Hunter College High School, *Annals*, January 1948.

38. Hunter College High School, *Annals*.

39. Dresselhaus, "Hunter College High School Address," 2009.

CHAPTER 3

1. "Leo Szilard," Atomic Heritage Foundation, https://www.atomi
cheritage.org/profile/leo-szilard.

2. Timothy J. Jorgensen, "Lise Meitner—The Forgotten Woman of
Nuclear Physics Who Deserved a Nobel Prize," *The Conversation*, February 7, 2019, http://theconversation.com/lise-meitner-the-forgotten
-woman-of-nuclear-physics-who-deserved-a-nobel-prize-106220; Ruth
H. Howes and Caroline C. Herzenberg, *Their Day in the Sun: Women
of the Manhattan Project* (Philadelphia: Temple University Press, 2003);
Leonore Tiefer, "How the Quad Went Coed," *Wall Street Journal*, November 21, 2016, https://www.wsj.com/articles/how-the-quad-went-coed
-1479680187.

3. "Hunter College Mission," Hunter College, https://hunter.cuny.edu
/about/mission/; Maura King, Hunter College Office of Legal Affairs,
email to author, September 4, 2018; Mildred Dresselhaus, interview
by Shirlee Sherkow, 1976, transcript, Project on Women as Scientists
and Engineers, MIT Libraries Distinctive Collections, Cambridge, MA,
29–30.

4. Mildred Dresselhaus, interview, 1976, 33–34.

5. Mildred Dresselhaus, interview, 1976, 35.

6. Mildred Dresselhaus, interview, 1976, 13.

7. Mildred Dresselhaus, interview, 1976, 37–38.

8. Maura King, email to author, September 4, 2018; Mildred Dresselhaus, "Expanding the Audience for Physics Education," presentation abstract from the American Association of Physics Teachers 2008
Winter Meeting, https://www.aapt.org/AbstractSearch/FullAbstract.cfm
?KeyID=15107.

9. Mildred Dresselhaus, interview by the US Department of Energy Office of Science, "Fermi Award Winners: Q&A," US Department of Energy Office of Science, June 6, 2012, https://web.archive.org/web/20150908034322/https://science.energy.gov/news/featured-articles/2012/06-06-12/; Mildred Dresselhaus, interview, 1976, 38.

10. Mildred Dresselhaus, interview, 1976, 39.

11. Mildred Dresselhaus, "Mildred Dresselhaus Biography," Kavli Prize, http://kavliprize.org/sites/default/files/%25nid%25/autobiagraphies_attachments/Mildred_Dresselhaus_Biography_0.pdf.

12. Sharon Bertsch McGrayne, *Nobel Prize Women in Science: Their Lives, Struggles, and Momentous Discoveries*, 2nd ed. (Washington, DC: Joseph Henry Press, 1998), 332–354, 93–116.

13. Mildred Dresselhaus, interview by Magdolna Hargittai, *Candid Science IV: Conversations with Famous Physicists* (London: Imperial College Press, 2004), 548.

14. Ruth H. Howes, "Rosalyn Sussman Yalow (1921–2011)," *Physics and Society*, American Physical Society, October 2001, https://www.aps.org/units/tps/newsletters/201110/howes.cfm; Eugene Straus, *Rosalyn Yalow Nobel Laureate: Her Life and Work in Medicine* (New York: Plenum Press, 1998), 33–34, 65–69; Mildred Dresselhaus and F. A. Stahl, "Rosalyn Sussman Yalow (1921–)," in *Out of the Shadows: Contributions of Twentieth-Century Women to Physics*, ed. Nina Byers and Gary Williams (Cambridge: Cambridge University Press, 2006), 307–308.

15. Straus, *Rosalyn Yalow Nobel Laureate*, 66.

16. Mildred Dresselhaus, interview, 1976, 29; Howes, "Rosalyn Sussman Yalow (1921–2011)."

17. Straus, *Rosalyn Yalow Nobel Laureate*, 66.

18. Mildred Dresselhaus, interview by Harry Kroto, Vega Science Trust, 2001, http://www.vega.org.uk/video/programme/20.

19. Mildred Dresselhaus, interview by US Department of Energy Office of Science, https://web.archive.org/web/20150908034322/https://science.energy.gov/news/featured-articles/2012/06-06-12/.

20. "History," Department of Physics, Columbia University, https://physics.columbia.edu/content/history; Straus, *Rosalyn Yalow Nobel Laureate*, 66–67.

21. Mildred Dresselhaus, interview by the Kavli Foundation, "2012 Kavli Prize in Nanoscience: A Discussion with Mildred Dresselhaus," August 2012, https://www.kavlifoundation.org/science-spotlights/kavli-prize-2012-dresselhaus#.XjEWVlBOnOR.

22. Natalie Angier, "Carbon Catalyst for Half a Century," *New York Times*, July 2, 2012, https://www.nytimes.com/2012/07/03/science/carbon-catalyst-for-half-a-century.html.

23. Mildred Dresselhaus, interview, 2004, 549.

24. Straus, *Rosalyn Yalow Nobel Laureate*, 77.

25. Mildred Dresselhaus, interview, 2004, 549.

26. Mildred Dresselhaus, interview, 1976, 36–37; "Mildred Dresselhaus," *Arlington Public News*, YouTube, February 5, 2015, https://youtu.be/0JOlyDyUYnw.

27. McGrayne, *Nobel Prize Women in Science*, 341; Mildred Dresselhaus, interview, 1976, 13.

28. Mildred Dresselhaus, interview, 1976, 39–40.

29. Mildred Dresselhaus, interview, 1976, 40–41, 56, 68.

30. Mildred Dresselhaus, interview, 1976, 53.

31. Mark Anderson, "The Queen of Carbon," *IEEE Spectrum* 52, no. 5 (2015): 52; Mildred Dresselhaus, interview, 1976, 56.

32. "Hunter College of the City of New York Commencement Exercises," Hunter College, June 21, 1951, https://library.hunter.cuny.edu/old/sites/default/files/100th_commencement_06211951.pdf; Marianne Dresselhaus Cooper, email to author, June 8, 2020.

33. Mildred Dresselhaus, interview, 1976, 42.

34. "Hunter College Commencement."

35. Mildred Dresselhaus, interview, 1976, 113–114.

36. Mildred Dresselhaus, interview, 1976, 58.

37. "Nobel Laureates," University of Cambridge Department of Physics, Cavendish Laboratory, https://www.phy.cam.ac.uk/history/nobel; Mildred Dresselhaus, interview, 1976, 57–59.

38. Mildred Dresselhaus, interview, 1976, 57.

39. Mildred Dresselhaus, interview, 1976, 62.

40. Shoshi Dresselhaus-Cooper, email to author, October 18, 2018; Anthony Tucker, "Sir Brian Pippard," *Guardian*, September 23, 2008, https://www.theguardian.com/science/2008/sep/24/physics.peoplein science; Paul Preuss, "Superconductors Face the Future," Lawrence Berkeley National Laboratory, September 10, 2010, https://newscenter .lbl.gov/2010/09/10/superconductors-future; Bridget Cunningham, "Mildred Dresselhaus, a Driving Force for Women in STEM," COM-SOL blog, March 7, 2016, https://www.comsol.com/blogs/mildred -dresselhaus-a-driving-force-for-women-in-stem/.

41. Mildred Dresselhaus, interview, 1976; Michael Berry and John Cornwell, "Robert Balson Dingle," Royal Society of Edinburgh, https.// www.rse.org.uk/cms/files/fellows/obits_alpha/dingle_robert.pdf; "Professor Bob Chambers, 1924–2016," University of Bristol, January 20, 2017, http://www.bristol.ac.uk/news/2017/january/bob-chambers.html; Charles Clement, "Tony Lane Obituary," *Guardian*, March 10, 2011, https://www.theguardian.com/science/2011/mar/10/tony-lane-obituary.

42. Mildred Dresselhaus, interview, 1976, 58–65.

43. Mildred Dresselhaus, interview, 1976, 63.

44. Mildred Dresselhaus, interview, 1976, 42.

45. Mildred Dresselhaus, interview, 1976, 65–68, Marianne Dresselhaus Cooper, interview by author, Arlington, MA, April 27, 2018.

46. Mildred Dresselhaus, interview, 1976, 67.

47. Mildred Dresselhaus, interview, 1976, 63.

48. Mildred Dresselhaus, interview, 1976, 66.

49. Mildred Dresselhaus, interview, 1976, 67.

50. Dorothy Elia Howells, *A Century to Celebrate: Radcliffe College, 1879–1979* (Cambridge, MA: Radcliffe College, 1978), 1–15; "Albert Einstein—Facts," NobelPrize.org, Nobel Media, https://www.nobelprize .org/prizes/physics/1921/einstein/facts/; Drew Gilpin Faust, "Mingling Promiscuously: A History of Women and Men at Harvard," in *Yards and Gates: Gender in Harvard and Radcliffe History*, ed. Laurel Ulrich (New York: Palgrave Macmillan, 2004), 317–328; Colleen Walsh, "Hard-Earned Gains for Women at Harvard," *Harvard Gazette*, April 26, 2012, https://news.harvard.edu/gazette/story/2012/04/hard -earned-gains-for-women-at-harvard/.

51. Walsh, "Hard-Earned Gains."

52. Howells, *A Century to Celebrate*, 14; Faust, "Mingling Promiscuously," 317.

53. Anderson, "Queen of Carbon," 53; Mildred Dresselhaus, interview, 1976.

54. Anderson, "Queen of Carbon," 53; Mildred Dresselhaus, interview by Martha A. Cotter and Mary S. Hartman, *Talking Leadership: Conversations with Powerful Women* (New Brunswick, NJ: Rutgers University Press, 1999), 70.

55. Mildred Dresselhaus, interview, 1999, 70.

56. Mildred Dresselhaus, interview, 1976, 32–33, 59–61.

57. Mildred Dresselhaus, interview, 1976, 117.

58. Mildred Dresselhaus, interview, 1976, 114–116.

59. Sam Merrill, "Women in Engineering," *Cosmopolitan*, April 1976, 162–166.

60. "Ruth Bader Ginsburg," Oyez, https://www.oyez.org/justices/ruth _bader_ginsburg; *RBG*, directed by Betsy West and Julie Cohen (Magnolia Pictures, 2018), https://www.amazon.com/gp/video/detail/B07 CT9Q5C6.

61. Anderson, "Queen of Carbon," 53; "Physicist Enrico Fermi Pro-
duces the First Nuclear Chain Reaction," History.com, A&E Television
Networks, November 16, 2009, https://www.history.com/this-day-in
-history/fermi-produces-the-first-nuclear-chain-reaction.

62. Anderson, "Queen of Carbon," 53.

CHAPTER 4

1. Steve Koppes, "How the First Chain Reaction Changed Science,"
University of Chicago, https://www.uchicago.edu/features/how_the
_first_chain_reaction_changed_science/; Ingred Gonçalves and Mau-
reen Searcy, "Manhattan's Critical Moment," *University of Chicago
Magazine* (Fall 2017): 56–57.

2. Koppes, "Chain Reaction"; Gonçalves and Searcy, "Critical
Moment," 56–57; "Manhattan Project Spotlight: Enrico Fermi," Atomic
Heritage Foundation, October 28, 2015, https://www.atomicheritage
.org/article/manhattan-project-spotlight-enrico-fermi; J. A. J. Gowlett,
"The Discovery of Fire by Humans: A Long and Convoluted Process,"
Philosophical Transactions of the Royal Society B 371, no. 1696 (June 5,
2016), http://doi.org/10.1098/rstb.2015.0164.

3. "Mildred Dresselhaus Biography," Kavli Prize, http://kavliprize.org
/sites/default/files/%25nid%25/autobiagraphies_attachments/Mildred
_Dresselhaus_Biography_0.pdf; Sharon Bertsch McGrayne, *Nobel
Prize Women in Science: Their Lives, Struggles and Momentous Discoveries*
(Washington, DC: Joseph Henry Press, 1998), 189–199; Ruth H. Howes
and Caroline L. Herzenberg, *Their Day in the Sun: Women of the Man-
hattan Project* (Philadelphia: Temple University Press, 1999), 192–193;
Mildred Dresselhaus, interview by Shirlee Sherkow, 1976, transcript,
Project on Women as Scientists and Engineers, MIT Libraries Distinc-
tive Collections, Cambridge, MA, 77.

4. Mark Anderson, "The Queen of Carbon," *IEEE Spectrum* 52, no. 5
(2015): 50–54; Franklin Institute, "Mildred Dresselhaus—2017 Laureate of the Franklin Institute in Materials Science and Engineering,"
YouTube, May 4, 2017, https://youtu.be/caCAPTIZtkY; Mildred Dresselhaus, interview, 1976, 62.

5. Natalie Angier, "Mildred Dresselhaus, Who Pioneered Revolution in Carbon Use, Dies at 86," *New York Times*, February 24, 2017, B-15.

6. Atomic Heritage Foundation, "Enrico Fermi."

7. Mildred Dresselhaus, interview by Harry Kroto, Vega Science Trust, 2001, http://www.vega.org.uk/video/programme/20.

8. Mildred Dresselhaus, interview, 2001.

9. Mildred Dresselhaus, interview, 2001.

10. Mildred Dresselhaus, interview by the Kavli Foundation, "2012 Kavli Prize in Nanoscience: A Discussion with Mildred Dresselhaus," August 2012, https://www.kavlifoundation.org/science-spotlights/kavli -prize-2012-dresselhaus#.XjEWVlBOnOR; Robbie Gonzalez, "Answer Quickly: How Many Piano Tuners Are There in the City of Chicago?" *Gizmodo*, September 12, 2012, https://io9.gizmodo.com/answer-quickly -how-many-piano-tuners-are-there-in-the-5942673.

11. "Enrico Fermi Awards Ceremony for Dr. Mildred S. Dresselhaus and Dr. Burton Richter, May 2012," US Department of Energy Office of Science and Technical Information, May 7, 2012, https://www.osti .gov/sciencecinema/biblio/1044165.

12. US Department of Energy, "Fermi Awards Ceremony"; Gonzalez, "Answer Quickly."

13. Kavli Prize, "Mildred Dresselhaus Biography."

14. Mildred Dresselhaus, interview, August 2012.

15. Jay Orear, "My First Meetings with Fermi," in *Fermi Remembered*, ed. James W. Cronin (Chicago: University of Chicago Press, 2004), 202–203; Louis Hand and Donald Holcomb, "Jay Orear," Cornell University, http://archive.theuniversityfaculty.cornell.edu/memorials /OREAR.pdf.

16. Greg Wientjes, *Creative Genius in Technology: Mentor Principles from Life Stories of Geniuses and Visionaries of the Singularity* (CreateSpace, 2011), 111.

17. Mildred Dresselhaus, interview by Brian Keegan, August 27, 2007, transcript, MIT Infinite History, Cambridge, MA, https://infinitehistory.mit.edu/video/mildred-s-dresselhaus.

18. Alice Dragoon, "The 'What If?' Whiz," *MIT Technology Review*, April 23, 2013, https://www.technologyreview.com/s/513491/the-what-if-whiz/.

19. Harold Agnew, "A Snapshot of My Interactions with Fermi," in *Fermi Remembered*, ed. James W. Cronin (Chicago: University of Chicago Press, 2004), 185; "Harold M. Agnew," Los Alamos National Laboratory, https://www.lanl.gov/about/history-innovation/lab-directors/harold-agnew.php.

20. US Department of Energy, "Fermi Awards Ceremony."

21. Mildred Dresselhaus, interview, August 2012.

22. Richard L. Garwin, "Enrico Fermi and Ethical Problems in Scientific Research," Federation of American Scientists, October 19, 2001, https://fas.org/rlg/011019-fermi.htm; Mildred Dresselhaus, interview, August 2012.

23. Mildred Dresselhaus, interview, 2001.

24. Mildred Dresselhaus, interview, August 2012.

25. Mildred Dresselhaus, interview, August 2012.

26. Mildred Dresselhaus, interview, August 2012.

27. "Our History," University of Chicago Department of Physics, https://physics.uchicago.edu/about/our-history; R. W. Keyes, "Andrew Lawson," *Physics Today* 31, no. 6 (1978): 69.

28. Mildred Dresselhaus, interview by Magdolna Hargittai, *Candid Science IV: Conversations with Famous Physicists* (London: Imperial College Press, 2004), 549–550.

29. Mildred Dresselhaus, interview, 2004.

30. Mildred Dresselhaus, interview, August 2012.

31. Sam Merrill, "Women in Engineering," *Cosmopolitan*, April 1976, 162.

32. Mildred Dresselhaus, interview, 1976, 69.

33. Mildred Dresselhaus, interview by Martha A. Cotter and Mary S. Hartman, *Talking Leadership: Conversations with Powerful Women* (New Brunswick, NJ: Rutgers University Press, 1999), 70–71.

34. Mildred Dresselhaus, interview, 1976, 72–73.

35. Lolly Boween, "Clyde A. Hutchison," *Chicago Tribune*, September 12, 2005, https://www.chicagotribune.com/news/ct-xpm-2005-09-12 -0509120161-story.html.

36. Mildred Dresselhaus, interview, August 2012.

37. Marianne Dresselhaus Cooper, interview by author, Arlington, MA, April 27, 2018; Eliot Dresselhaus, web video interview by author, April 29, 2020; Editors of Encyclopaedia Britannica, "Canal Zone," *Encyclopaedia Britannica*, Encyclopaedia Britannica, May 1, 2020, https://www .britannica.com/place/Canal-Zone; "Charles Kittel," American Institute of Physics, https://history.aip.org/phn/11505002.html.

38. Gene Dresselhaus, A. F. Kip, and C. Kittel, "Plasma Resonance in Crystals: Observations and Theory," *Physical Review* 100 (October 15, 1955): 618; Gene Dresselhaus, A. F. Kip, and C. Kittel, "Cyclotron Resonance of Electrons and Holes in Silicon and Germanium Crystals," *Physical Review* 98 (April 15, 1955): 368; Gene Dresselhaus, "Spin-Orbit Coupling Effects in Zinc Blende Structures," *Physical Review* 100 (October 15, 1955): 580; Paul Dresselhaus, email to author, May 26, 2020; Wan-Tsang Wang et al., "Dresselhaus Effect in Bulk Wurtzite Materials," *Applied Physics Letters* 91, no. 8 (August 24, 2007), https://doi.org /10.1063/1.2775038.

39. Paul Dresselhaus, email to author, November 12, 2019.

40. Gene Dresselhaus, "Spin-Orbit Coupling." Shoshi Dresselhaus-Cooper, email to author, May 16, 2018.

41. "Remembering Frederick Reif," University of California at Berkeley Department of Physics, August 26, 2019, https://physics.berkeley.edu /news-events/news/20190826/remembering-frederick-reif; "Mildre[d] Reif," New York State Marriage Index, 1881–1967, New York State

Department of Health, Albany, NY, Ancestry.com; Erica Lehrer, email to author, May 29, 2020; Sherwin Lehrer, email to author, June 9, 2020.

42. Mildred Dresselhaus, interview, 1976, 69.

43. Shoshi Dresselhaus-Cooper, email to author, May 16, 2018.

44. Shoshi Dresselhaus-Cooper, email, May 16, 2018.

45. "The Nobel Prize in Physics 1972," NobelPrize.org, Nobel Media AB 2020, https://www.nobelprize.org/prizes/physics/1972/summary; CERN, "Superconductivity"; https://home.cern/science/engineering /superconductivity; Michael Sutherland, "Explainer: What Is a Super-conductor?" *The Conversation*, March 4, 2015, https://theconversation .com/explainer-what-is-a-superconductor-38122; Mildred Dresselhaus, interview, August 2012.

46. US Department of Energy, "Fermi Awards Ceremony."

47. Anderson, "The Queen of Carbon," 53.

48. Mildred Dresselhaus, interview by Bernadette Bensaude-Vincent and Arne Hessenbruch, October 25, 2001, transcript, History of Recent Science and Technology Project, Dibner Institute for the History of Science and Technology at MIT, Cambridge, MA, https://authors .library.caltech.edu/5456/1/hrst.mit.edu/hrs/materials/public/Dressel haus/Dresselhaus(HelenaFu_plus).html.

49. CERN, "Superconductivity"; Mildred Dresselhaus, interview, October 2001.

50. Natalie Angier, "Carbon Catalyst for Half a Century," *New York Times*, July 2, 2012, https://www.nytimes.com/2012/07/03/science /carbon-catalyst-for-half-a-century.html.

51. Mildred Dresselhaus, interview, October 2001.

52. Editors of Encyclopaedia Britannica, "BCS Theory," *Encyclopaedia Britannica*, May 30, 2017, https://www.britannica.com/science/BCS -theory; Adam Mann, "High-Temperature Superconductivity at 25: Still in Suspense," *Nature* 475 (July 20, 2011): 280–282, https://doi.org /10.1038/475280a.

53. Kavli Prize, "Mildred Dresselhaus Biography"; Mildred Dressel-haus, interview with Joseph D. Martin, transcript, American Institute of Physics, Niels Bohr Library and Archive, College Park, MD, June 24, 2014; Mildred Dresselhaus, interview, August 2012.

54. Mildred Dresselhaus, interview, October 2001.

55. Mildred Dresselhaus, interview, August 2012; "John Bardeen—Biographical," NobelPrize.org, Nobel Media AB 2020, https://www.nobelprize.org/prizes/physics/1956/bardeen/biographical.

56. Mildred Dresselhaus, interview, October 2001.

57. Press release, NobelPrize.org, Nobel Media AB 2020, October 20, 1972, https://www.nobelprize.org/prizes/physics/1972/press-release.

58. Mildred Dresselhaus, interview, 2001.

59. Gene Dresselhaus and Mildred Dresselhaus, foreword to *Anomalous Effects in Simple Metals* by Albert Overhauser (Weinheim: Wiley-VCH, 2011), vi; "Mildred Dresselhaus," Kavli Prize.

60. Mildred Dresselhaus, interview by the US Department of Energy Office of Science, "Fermi Award Winners: Q&A," US Department of Energy Office of Science, June 6, 2012, https://web.archive.org/web/20150908034322/https://science.energy.gov/news/featured-articles/2012/06-06-12/.

61. Mildred Dresselhaus, interview, August 2012.

62. Marianne Dresselhaus Cooper, email to author, June 8, 2020; Mildred Dresselhaus, interview by Paul S. Weiss, "A Conversation with Prof. Mildred Dresselhaus: A Career in Carbon Nanomaterials," *ACS Nano* 3, no. 9 (September 2009): 2434–2440.

63. JoAnn Creviston, University of Chicago Office of the Registrar, email to author, October 30, 2018; Mildred Dresselhaus, interview, 2004.

64. Mildred Dresselhaus, interview, 1976, 141–142; Mildred Spiewak, "Magnetic Field Dependence of High-Frequency Penetration into a

Superconductor," *Physical Review Letters* 1 (August 1958): 136; Mildred Spiewak, "Magnetic Field Dependence of the Surface Impedance of Superconducting Tin," *Physical Review* 113 (March 1959): 1479; Mildred Dresselhaus, curriculum vitae, unpublished.

65. Mildred Dresselhaus, interview, 2009, 2437.

66. G. Dresselhaus and M. Dresselhaus, foreword, vi.

67. G. Dresselhaus and M. Dresselhaus, foreword, vi.

68. Lynnette D. Madsen, *Successful Women Ceramic and Glass Scientists and Engineers: 100 Inspirational Profiles* (Hoboken, NJ: Wiley, 2016), 122.

69. Mildred Dresselhaus, interview, 2009, 2437.

70. Mildred Dresselhaus, interview, 2004, 551–552.

71. Mildred Dresselhaus, interview, 2004, 551–552.

72. Mildred Dresselhaus, interview, 2004, 551–552.

73. Marianne Dresselhaus Cooper, "My Extended Family: Growing Up as the Daughter of Millie Dresselhaus," Celebrating Millie, May 15, 2018, https://millie.pubpub.org/pub/6c8d1jyi.

74. Laura Doughty, interview by author, Wendell, MA, October 10, 2019.

75. Mildred Dresselhaus, interview, 2009, 2437.

76. Mildred Dresselhaus, interview, 2009, 2437.

77. Mildred Dresselhaus, interview, 1976, 142–143.

78. Marianne Dresselhaus Cooper, interview, 2018.

79. Elizabeth Dresselhaus, email to author, November 15, 2019.

80. Doughty, interview, 2019.

81. Mildred Dresselhaus, interview, 1976, 144–146.

82. Mildred Dresselhaus, interview, 1976, 150–151.

83. Millie and Gene were offered jobs at IBM's Manhattan facility next to Columbia University and would have moved to the company's current Yorktown Heights location if they had accepted. Mildred Dresselhaus, interview, 2009, 2437; Emerson W. Pugh, *Building IBM: Shaping an Industry and Its Technology* (Cambridge, MA: MIT Press, 1995), 127–128, 229, 237.

84. Pugh, *Building IBM*, 237–240.

85. Mildred Dresselhaus, interview, 2009, 2437; Mildred Dresselhaus, interview, 1976, 146–151.

86. Mildred Dresselhaus, interview, 1976, 146.

87. "About," Lincoln Laboratory, MA Institute of Technology, https://www.ll.mit.edu/about; "SAGE: Semi-Automatic Ground Environment Air Defense System," Lincoln Laboratory, MIT, https://www.ll.mit.edu/about/history/sage-semi-automatic-ground-environment-air-defense-system.

88. Lincoln Laboratory, "SAGE"; "MIT Radiation Laboratory," Lincoln Laboratory, MIT, https://www.ll.mit.edu/about/history/mit-radiation-laboratory.

89. Lincoln Laboratory, "Sage"; "Sputnik 1," NASA, October 4, 2011, https://www.nasa.gov/multimedia/imagegallery/image_feature_924.html.

90. Mildred Dresselhaus, interview, 1976, 144–146; Roshan L. Aggarwal and Marion B. Reine, "Benjamin Lax 1915–2015," Biographical Memoirs, National Academy of Sciences (2016), 6–7, http://www.nasonline.org/publications/biographical-memoirs/memoir-pdfs/lax-benjamin.pdf; Mildred Dresselhaus, interview, 2004, 551.

91. Mildred Dresselhaus, interview, 2007.

92. Mildred Dresselhaus, interview, 1976, 160–162; "Mildred Dresselhaus," *Arlington Public News*, YouTube, February 5, 2015, https://youtu.be/0JOlyDyUYnw.

93. "Friends of Menotomy Rocks Park," https://friendsofmenotomy.org.

CHAPTER 5

1. Roshan L. Aggarwal and Marion B. Reine, "Benjamin Lax 1915–2015," Biographical Memoirs, National Academy of Sciences (2016), 1–4, http://www.nasonline.org/publications/biographical-memoirs/memoir-pdfs/lax-benjamin.pdf; "Plasma," Encyclopaedia Britannica, Encyclopaedia Britannica, Inc., July 18, 2019, https://www.britannica.com/science/plasma-state-of-matter.

2. Aggarwal and Reine, "Benjamin Lax," 6–7; William Coffeen Holton and S. M. Sze, "Semiconductor Device," *Encyclopaedia Britannica*, April 10, 2016, https://www.britannica.com/technology/semiconductor-device.

3. "Semimetal," Lexico.com, https://www.lexico.com/definition/semimetal; Aggarwal and Reine, "Benjamin Lax," 6–7.

4. Aggarwal and Reine, "Benjamin Lax," 7–9.

5. Aggarwal and Reine, "Benjamin Lax," 7–9; L. G. Rubin, R. J. Weggel, E. J. McNiff Jr., and T. Vu, "The Francis Bitter National Magnet Laboratory at MIT: An Update," *Physica B: Condensed Matter* 201 (July–August 1994): 500; Marion Reine, email to author, May 3, 2020.

6. Benjamin Lax, interview by Donald T. Stevenson, *Benjamin Lax—Interviews on a Life in Physics at MIT* (Boca Raton, FL: CRC Press, 2020), 136–140; Mildred Dresselhaus, interview by Brian Keegan, August 27, 2007, transcript, MIT Infinite History, Cambridge, MA, https://infinitehistory.mit.edu/video/mildred-s-dresselhaus; Paul Dresselhaus, email to author, May 26, 2020.

7. Benjamin Lax, interview, 2020, 113; "Laura M. Roth," MIT Museum, https://webmuseum.mit.edu/detail.php?module=people&type=related&kv=17132; Mildred Dresselhaus, interview by Shirlee Sherkow, 1976, transcript, Project on Women as Scientists and Engineers, MIT Libraries Distinctive Collections, Cambridge, MA, 144–145.

8. Mildred Dresselhaus, interview, 2007.

9. Mildred Dresselhaus, interview by the Kavli Foundation, "2012 Kavli Prize in Nanoscience: A Discussion with Mildred Dresselhaus,"

August 2012, https://www.kavlifoundation.org/science-spotlights/kavli-prize-2012-dresselhaus#.XjEWVlBOnOR.

10. Jeff Hecht, "Short History of Laser Development," *Optical Engineering* 49, no. 2 (September 1, 2010): 091002, https://doi.org/10.1117/1.3483597.

11. Mildred Dresselhaus, interview, 2007.

12. Mildred Dresselhaus, interview, 2007.

13. "Magneto-optic." Merriam Webster.com Dictionary, Merriam-Webster, https://www.merriam-webster.com/dictionary/magneto-optic; Mildred Dresselhaus, interview by Harry Kroto, Vega Science Trust, 2001, http://www.vega.org.uk/video/programme/20.

14. Mildred Dresselhaus, interview, August 2012.

15. Mildred Dresselhaus, interview, August 2012; J.M.K.C. Donev et al., "Valence Band," Energy Education, 2018, https://energyeducation.ca/encyclopedia/Valence_band; J. M. K. C. Donev et al., "Conduction Band," Energy Education, 2018, https://energyeducation.ca/encyclopedia/Conduction_band, Mildred Dresselhaus, interview by Magdolna Hargittai, *Candid Science IV: Conversations with Famous Physicists* (London: Imperial College Press, 2004), 561.

16. Mildred Dresselhaus, interview, 2004, 561.

17. Mildred Dresselhaus, interview, 2004, 561; Mildred Dresselhaus, interview, 1976, 170.

18. John Emsley, "Bismuth," Education in Chemistry, Royal Society of Chemistry, November 19, 2014, https://edu.rsc.org/elements/bismuth/2000017.article; Mildred Dresselhaus, interview, 1976, 170–171.

19. Mildred Dresselhaus, interview, 1976, 171–177.

20. Mildred Dresselhaus, interview, 1976, 177. According to Millie, the individual had issues with others as well and ended up leaving MIT shortly after the clash with her. Once this person had gone, Millie returned to bismuth, then and many other times in her career.

21. Mildred Dresselhaus, interview, 2001.

22. Jeanie Chung, "Superconductor," *University of Chicago Magazine*, Summer 2015, https://mag.uchicago.edu/science-medicine/supercon ductor.

23. Mildred Dresselhaus, interview by Bernadette Bensaude-Vincent and Arne Hessenbruch, October 25, 2001, transcript, History of Recent Science and Technology Project, Dibner Institute for the History of Science and Technology at MIT, Cambridge, MA, https://ethw.org /Oral-History:Mildred_Dresselhaus; Mildred Dresselhaus, interview by Paul S. Weiss, "A Conversation with Prof. Mildred Dresselhaus: A Career in Carbon Nanomaterials," *ACS Nano* 3, no. 9 (September 2009): 2434–2440.

24. Chung, "Superconductor."

25. Natalie Angier, "Carbon Catalyst for Half a Century," *New York Times*, July 2, 2012, https://www.nytimes.com/2012/07/03/science /carbon-catalyst-for-half-a-century.html.

26. Mildred Dresselhaus, interview, October 2001; Mildred Dressel-haus, interview, September 2009.

27. Mildred Dresselhaus, interview, 2004, 561.

28. Mildred Dresselhaus, interview, August 2012; Reine, email, 2020.

29. Mildred Dresselhaus, interview, 2001.

30. Mildred Dresselhaus, interview, August 2009.

31. Mildred Dresselhaus, interview, 2004, 561.

32. Paul Dresselhaus, "Growing Up with Millie," speech, MIT, Cam-bridge, MA, Nov. 26, 2017, http://web.mit.edu/webcast/millie; Paul Dresselhaus, email to author, May 26, 2020.

33. Paul Dresselhaus, email to author, November 12, 2019.

34. Mildred Dresselhaus, interview, 1976, 147–149;

35. Angier, "Carbon Catalyst."

36. Mildred Dresselhaus, interview by Vijaysree Venkatraman, "Reflec-tions of a Woman Pioneer," *Science*, November 11, 2014, https://

www.sciencemag.org/careers/2014/11/reflections-woman-pioneer; Alice Dragoon, "The 'What If?' Whiz," *MIT Technology Review*, April 23, 2013, https://www.technologyreview.com/s/513491/the-what-if-whiz/.

37. Mildred Dresselhaus, interview, 1976, 159.

38. Mildred Dresselhaus, interview, 1976, 159.

39. Natalie Angier, "Mildred Dresselhaus, Who Pioneered Revolution in Carbon Use, Dies at 86," *New York Times*, February 24, 2017, B-15; "Women as a Percentage of Total Undergraduates, Graduate Students, and Faculty: Academic Years 1901–2014," *MIT Faculty Newsletter 26*, vol. 4 (March/April 2014), http://wcb.mit.edu/fnl/volume/264/numbers.html.

40. Angier, "Mildred Dresselhaus . . . Dies at 86."

41. Mildred Dresselhaus, interview by the US Department of Energy Office of Science, "Fermi Award Winners: Q&A," US Department of Energy Office of Science, June 6, 2012, https://web.archive.org/web/20150908034322/https://science.energy.gov/news/featured-articles/2012/06-06-12/.

42. Angier, "Carbon Catalyst"; Mildred Dresselhaus, interview, 2007.

43. Mildred Dresselhaus, interview, 2007.

44. Paul Coxon, "Have Scientists Really Found Something Harder Than Diamond?" *Conversation*, January 19, 2016, https://theconversation.com/have-scientists-really-found-something-harder-than-diamond-52391; Seth I. Rosen, "Are Diamonds Really Rare? Diamond Myths and Misconceptions," International Gem Society, https://www.gemsociety.org/article/are-diamonds-really-rare.

45. "The Unleaded Pencil," Pencils.com, https://pencils.com/pages/the-unleaded-pencil; D. D. Richardson, "A Calculation of Van der Waals Interactions in and between Layers of Atoms: Application to Graphite," *Journal of Physics C: Solid State Physics* 10 (1977): 3235; Ethan Siegel, "There Are 6 'Strongest Materials' on Earth That Are Harder Than Diamonds," *Forbes*, June 18, 2019, https://www.forbes.com/sites/startswithabang/2019/06/18/there-are-6-strongest-materials-on-earth

-that-are-harder-than-diamonds/#3e5785cd3412; "Fillo, Filo, or Phyllo?" Fillo Factory, https://www.fillofactory.com/phyllo-dough-s/122.htm.

46. "The Unleaded Pencil"; Hobart M. King, "Graphite," Geology. com, https://geology.com/minerals/graphite.shtml.

47. There are dozens of examples online for making pencil-based circuits, using either full pencils or pencil-drawn circuits, such as "Graphite Circuit," KiwiCo, https://www.kiwico.com/diy/Science-Projects -for-Kids/3/project/Graphite-Circuit/2667; "The Nobel Prize in Physics 2010," NobelPrize.org, Nobel Media AB 2020, https://www.nobelprize .org/prizes/physics/2010/summary.

48. "The Nobel Prize in Physics 2010."

49. Mildred Dresselhaus, interview, 2001; Benjamin Lax, interview, 2020, 135–140.

50. Mildred Dresselhaus, interview, 2007.

51. Mildred Dresselhaus, interview by Steve Yalisove, "Advancing Carbon, Energy Materials: Mildred S. Dresselhaus Talks about Her Work," *MRS Bulletin* 38, no. 11 (November 2013): 974–976.

52. Mildred Dresselhaus, interview, October 2001.

53. Mildred Dresselhaus, interview, October 2001; L C. F. Blackman and A. R. Ubbelohde, "Stress Recrystallization of Graphite," *Proceedings of the Royal Society of London* 266, no. 1324 (February 27, 1962), 20–32; "Dresselhaus Wins L'OREAL-UNESCO for Women in Science Prize," *SEED*, February 22, 2007, https://www.seedmagazine.com/content /article/dresselhaus_wins_loreal-unesco_for_women_in_science_prize.

54. Mildred Dresselhaus, interview, October 2001; Mildred Dresselhaus, interview, 2004, 561.

55. Mildred Dresselhaus, interview, 2001; Mildred Dresselhaus, interview, October 2001; Dragoon, "The 'What If?' Whiz."

56. Mildred Dresselhaus, interview, October 2001; "Joel McClure," *Eugene Register-Guard*, September 11, 2016, https://www.legacy.com /obituaries/RegisterGuard/obituary.aspx?page=lifestory&pid=18136 3887.

57. Mildred Dresselhaus, interview, October 2001.

58. Mildred S. Dresselhaus and John G. Mavroides, "The Fermi Surface of Graphite," *IBM Journal of Research and Development* 8, no. 3 (July 1964): 262–267; Joel W. McClure, "Energy Band Structure of Graphite," *IBM Journal of Research and Development* 8, no. 3 (July 1964): 255–261; Sidney Perkowitz, "Fermi Surface," *Encyclopaedia Britannica*, Encyclopaedia Britannica, Inc., June 14, 2013, https://www.britannica.com/science/Fermi-surface.

59. Dragoon, "The 'What If?' Whiz."

60. Mildred Dresselhaus, interview, 2004, 562.

61. OpenStax, "Atoms, Isotopes, Ions, and Molecules: The Building Blocks" in Biology, OpenStax CNX, May 8, 2019 http://cnx.org/contents/185cbf87-c72e-48f5-b51e-f14f21b5eabd@11.10.

62. Tony R. Kuphaldt, "Band Theory of Solids," in *Semiconductors*, All About Circuits editorial team, https://www.allaboutcircuits.com/textbook/semiconductors/chpt-2/band-theory-of-solids.

63. Kuphaldt, "Band Theory of Solids."

64. Donev et al., "Valence Band"; Donev et al., "Conduction Band."

65. Donev et al., "Conduction Band."

66. Donev et al., "Valence Band."

67. While pure diamond is an insulator, diamond can become a semiconductor when impurities are added. "Diamond Factory: Expert Q&A," *NOVA*, April 1, 2009, https://www.pbs.org/wgbh/nova/article/butler-diamonds.

68. Donev et al., "Conduction Band."

69. Lexico.com, "Semimetal"; Jeanie Chung, "Superconductor."

70. Mildred Dresselhaus, "Mildred Dresselhaus Biography," Kavli Prize, http://kavliprize.org/sites/default/files/%25nid%25/autobiagraphies_attachments/Mildred_Dresselhaus_Biography_0.pdf; Mildred Dresselhaus, curriculum vitae, unpublished; Mildred Dresselhaus, interview, 1976, 155.

71. Ephrat Livni and Dan Kopf, "The Decline of the Large US Family, in Charts," *Quartz*, October 11, 2017, https://qz.com/1099800/average-size-of-a-us-family-from-1850-to-the-present; Mildred Dresselhaus, interview, 2004, 556; Mildred Dresselhaus, interview, 1976, 177–180.

72. Mark Anderson, "The Queen of Carbon," *IEEE Spectrum* 52, no. 5 (May 2015): 54.

73. Benjamin Lax, interview, 2020, 171; Mildred Dresselhaus, interview, 1976, 170–171, 174, 188–190; Mildred Dresselhaus, interview, 2001; Joe Holley, "Biophysicist Samuel Williamson Dies," *Washington Post*, April 30, 2005, https://www.washingtonpost.com/wp-dyn/content/article/2005/04/29/AR2005042901621.html.

74. Benjamin Lax, interview, 2020, ix, 153.

75. Mildred Dresselhaus, interview, September 2009, 2437.

76. Mildred Dresselhaus, interview, 2004, 550.

77. Mildred Dresselhaus, interview, 1976, 178–180.

78. Mildred Dresselhaus, interview, 2004, 551.

79. "MIT History: The Women of the Institute Panel Discussion," MIT Infinite History, transcript, Cambridge, MA, June 24, 1997, https://infinitehistory.mit.edu/video/mit-history-women-institute-panel-discussion-6241997; Mildred Dresselhaus, interview, 1976, 180–182; Mildred Dresselhaus, interview, 2007.

80. Mildred Dresselhaus, interview, September 2009, 2437.

81. Mildred Dresselhaus, interview, 1976, 181–182, 199; Mildred Dresselhaus, interview, September 2009, 2437.

82. secooper87, "Mildred S. Dresselhaus' Retirement Party," YouTube video, 9:44, November 11, 2007, https://youtu.be/ixbbY4Jxx9Q.

83. Mildred Dresselhaus, interview, 1976, 181–182.

84. Mildred Dresselhaus, interview, 1976, 182–183; Margaret W. Rossiter, *Women Scientists in America: Before Affirmative Action 1940–1972*

(Baltimore: Johns Hopkins University Press, 1995), 38–39, 428; "Abby Rockefeller Mauze, Philanthropist, 72, Is Dead," *New York Times*, May 29, 1976, 26; "Endowed Professorships at MIT: A History," copy in the Office of the Provost, MIT, Cambridge, MA, June 1984; Ted Nygreen, "Dr. Hodgkin Serves as First Mauze Professor," *Tech*, November 17, 1965, http://tech.mit.edu/V85/PDF/V85-N24.pdf.

85. Mildred Dresselhaus, interview, September 2009, 2437; Mildred Dresselhaus, interview, 1976, 182; Mildred Dresselhaus, interview, 2007.

86. Mildred Dresselhaus, interview, 2004, 554.

CHAPTER 6

1. As baseball fans know, the World Series ended in Boston's defeat, as this was only about halfway through the Curse of the Bambino, an eighty-six-year drought between Red Sox World Series titles. *Boston Globe*, October 8, 1967, 210.

2. "Millie: Trailblazer for Women of MIT," Association of MIT Alumnae, Celebration of the Life of Millie Dresselhaus, November 26, 2017, http://amita.alumgroup.mit.edu/s/1314/images/gid20/editor_documents/millie/amita_celebrates_millie1.pdf; Sheila Widnall, "Millie's Impact on Women at MIT," prerecorded lecture, Celebrating Our Millie, MIT, Cambridge, MA, November 26, 2017.

3. Mark Anderson, "The Queen of Carbon," *IEEE Spectrum* 52, no. 5 (May 2015): 54.

4. Mildred Dresselhaus, interview by the Kavli Foundation, "2012 Kavli Prize in Nanoscience: A Discussion with Mildred Dresselhaus," August 2012, https://www.kavlifoundation.org/science-spotlights/kavli-prize-2012-dresselhaus#.XjEWVlBOnOR.

5. Joseph D. Martin, "Mildred Dresselhaus and Solid State Pedagogy at MIT," *Annalen der Physik* 2019, no. 531 (August 2019), https://doi.org/10.1002/andp.201900274.

6. Mildred Dresselhaus, interview by Bernadette Bensaude-Vincent and Arne Hessenbruch, October 25, 2001, transcript, History of Recent

Science and Technology Project, Dibner Institute for the History of Science and Technology at MIT, Cambridge, MA, https://ethw.org/Oral-History:Mildred_Dresselhaus.

7. Alice Dragoon, "The 'What If?' Whiz," *MIT Technology Review*, April 23, 2013, https://www.technologyreview.com/s/513491/the-what-if-whiz/.

8. Andrew Grant, "Mildred Dresselhaus (1930–2017)," *Physics Today*, February 23, 2017, https://physicstoday.scitation.org/do/10.1063/PT.5.9088/full/.

9. Mildred Dresselhaus, interview by Brian Keegan, August 27, 2007, transcript, MIT Infinite History, Cambridge, MA, https://infinitehistory.mit.edu/video/mildred-s-dresselhaus.

10. Mildred Dresselhaus, interview by Magdolna Hargittai, *Candid Science IV: Conversations with Famous Physicists* (London: Imperial College Press, 2004), 558; Margaret W. Rossiter, *Women Scientists in America: Before Affirmative Action, 1940–1972* (Baltimore: Johns Hopkins University Press, 1995), 428; Mildred Dresselhaus, interview by Shirlee Sherkow, 1976, transcript, Project on Women as Scientists and Engineers, MIT Libraries Distinctive Collections, Cambridge, MA, 184.

11. "Women as a Percentage of Total Undergraduates, Graduate Students, and Faculty: Academic Years 1901–2014," MIT Institutional Research, *MIT Faculty Newsletter* 26, vol. 4 (March/April 2014): 16; Lydia Snover, MIT Institutional Research, email to author, March 3, 2020; "Enrollments 2019–2020," in *MIT Facts 2020* (Cambridge, MA: MIT Reference Publications, 2020), 20; Catherine Hill, Christianne Corbett, and Andresse St. Rose, *Why So Few?: Women in Science, Technology, Engineering, and Mathematics* (Washington, DC: AAUW, 2010), 12–15; "Legal Highlight: The Civil Rights Act of 1964," US Department of Labor, https://www.dol.gov/agencies/oasam/civil-rights-center/statutes/civil-rights-act-of-1964; "Title IX and Sex Discrimination," US Department of Education, https://www2.ed.gov/about/offices/list/ocr/docs/tix_dis.html.

12. Mildred Dresselhaus, interview, 2004, 558.

13. Mildred Dresselhaus, interview, 2007.

14. Mildred Dresselhaus, interview by Paul S. Weiss, "A Conversation with Prof. Mildred Dresselhaus: A Career in Carbon Nanomaterials," *ACS Nano* 3, no. 9 (September 2009): 2438; Dragoon, "The 'What If?' Whiz."

15. "Years That Men's Colleges Became Co-ed," Collegexpress, https://www.collegexpress.com/lists/list/years-that-mens-colleges-became-co-ed/366. Some of these universities at that time educated women via affiliated colleges—Harvard via Radcliffe and Brown via Pembroke, for example.

16. Mildred Dresselhaus, interview, 2009, 2437–2438.

17. Aviva Brecher, "Remembering My Mentor, Millie," lecture, Celebrating Our Millie, MIT, Cambridge, MA, November 26, 2017.

18. "Institute Professor Emerita Mildred Dresselhaus, a Pioneer in the Electronic Properties of Materials, Dies at 86," *MIT News*, February 21, 2017, http://news.mit.edu/2017/institute-professor-emerita-mildred-dresselhaus-dies-86-0221; Widnall, "Millie's Impact on Women at MIT"; "Appointments—Jobs 1953 Forward," Emily Wick Papers, MC 696, box 1, MIT Libraries Distinctive Collections, Cambridge, MA; Stephen Salk, MIT Human Resources, email to author, November 12, 2020.

19. Sam Merrill, "Women in Engineering," *Cosmopolitan*, April 1976, 166.

20. "MRS Obituary for Arthur R. von Hippel," Materials Research Society, 2004, https://www.mrs.org/mrs-von-hippel-obituary; Mildred Dresselhaus, "Memories of Arthur von Hippel, 1898–2003," *MRS Bulletin* 39 (November 2014): 998–1003.

21. Dresselhaus, "Memories of Arthur von Hippel," 1000.

22. Mildred Dresselhaus, interview by Bernadette Bensaude-Vincent and Arne Hessenbruch, October 25, 2001, transcript, History of Recent Science and Technology Project, Dibner Institute for the History of

Science and Technology at MIT, Cambridge, MA, https://ethw.org /Oral-History:Mildred_Dresselhaus.

23. "Ali Javan, Scientist and Inventor—Obituary," *Telegraph*, September 21, 2016, https://www.telegraph.co.uk/obituaries/2016/09/21/ali -javan-scientist-and-inventor--obituary; Chuck Leddy, "Professor Emeritus Ali Javan, Inventor of the First Gas Laser, Dies at 89," *MIT News*, September 29, 2016, http://news.mit.edu/2016/physics-professor -emeritus-ali-javan-dies-0929.

24. Mildred Dresselhaus, curriculum vitae, unpublished; Mildred Dresselhaus, "New Materials through Science and New Science through Materials," James R. Killian Jr. Lecture, MIT, Cambridge, MA, April 8, 1987, https://youtu.be/ZOFHDo20YYc.

25. Mildred Dresselhaus, "New Materials through Science."

26. P. R. Schroeder, M. S. Dresselhaus, and A. Javan, "Location of Electron and Hole Carriers in Graphite from Laser Magnetoreflection Data," *Physical Review Letters* 20, no. 23 (June 3, 1968): 1292–1295; Mildred Dresselhaus, interview, 2004, 565.

27. Mildred Dresselhaus, "New Materials through Science."

28. Maia Weinstock, "Chien-Shiung Wu: Courageous Hero of Physics" in *A Passion for Science: Stories of Discovery and Invention*, ed. Suw Charman-Anderson (London: Finding Ada, 2013), chap. 16, https:// findingada.com/shop/a-passion-for-science-stories-of-discovery-and -invention.

29. Mildred Dresselhaus, interview, 2004, 565; J.M.K.C. Donev et al., "Charge Carrier," Energy Education, 2018, https://energyeducation.ca /encyclopedia/Charge_carrier.

30. J.M.K.C. Donev et al., "Electron Hole," Energy Education, 2018, https://energyeducation.ca/encyclopedia/Electron_hole.

31. Mildred Dresslhaus, interview, October 2001.

32. Mildred Dresselhaus, interview by Harry Kroto, Vega Science Trust, 2001, http://www.vega.org.uk/video/programme/20.

33. Mildred Dresselhaus, "New Materials through Science."

34. Mildred Dresselhaus, interview, 2001.

35. Mildred Dresselhaus, interview, 2001.

36. Ednah Dow Littlehale Cheney, *Memoir of Margaret Swan Cheney* (Boston: Lee and Shepard, 1889), 24–26; Sally Atwood, "A Haven for Women: One Alumna's Legacy," *MIT Technology Review*, September 1, 2005, https://www.technologyreview.com/s/404619/a-haven-for-women-one-alumnas-legacy; "Early Maps of the Massachusetts Institute of Technology," MIT Libraries, https://wayback.archive-it.org/7963/20190702002432/https://libraries.mit.edu/archives/exhibits/maps/index.html; "Margaret Cheney Room," MIT Division of Student Life, http://studentlife.mit.edu/impact-opportunities/diversity-inclusion/womenmit/margaret-cheney-room.

37. Amy Sue Bix, *Girls Coming to Tech!: A History of American Engineering Education for Women* (Cambridge, MA: MIT Press, 2013), 244; Atwood, "A Haven for Women."

38. Bix, *Girls Coming to Tech!*, 240–242.

39. MIT Institutional Research, "Women as a Percentage of Total Undergraduates, Graduate Students, and Faculty"; Bix, *Girls Coming to Tech!*, 230–253.

40. Widnall, "Millie's Impact on Women at MIT."

41. Bix, *Girls Coming to Tech!*, 245.

42. Mildred Dresselhaus, interview by Vijaysree Venkatraman, "Reflections of a Woman Pioneer," *Science*, November 11, 2014, https://www.sciencemag.org/careers/2014/11/reflections-woman-pioneer.

43. Mildred Dresselhaus, interview, 1976, 201–205.

44. Association of MIT Alumnae, "Millie: Trailblazer for Women of MIT."

45. Shirley M. Malcom, Janet Welsh Brown, and Paula Quick Hall, "The Double Bind: The Price of Being a Minority Woman in Science:

Report of a Conference of Minority Women Scientists" (Washington, DC: American Association for the Advancement of Science, 1976).

46. Shirley Ann Jackson, prerecorded lecture, Celebrating Our Millie, MIT, Cambridge, MA, Nov. 26, 2017.

47. "MIT History: The Women of the Institute Panel Discussion," MIT Infinite History, transcript, Cambridge, MA, June 24, 1997, https:// infinitehistory.mit.edu/video/mit-history-women-institute-panel -discussion-6241997.

48. Mildred Dresselhaus, interview, 1976, 203–206.

49. Elizabeth Durant, "Ellencyclopedia," *MIT Technology Review*, August 15, 2007, https://www.technologyreview.com/s/408456/ellen cyclopedia; Bix, *Girls Coming to Tech!*, 223–243; Mildred Dresselhaus, interview, 1976, 206–207.

50. Mildred Dresselhaus, interview, 2007.

51. Mildred Dresselhaus, interview, 1976, 206–208.

52. Emily Wick, "Proposal for a New Policy for Admission of Women Undergraduate Students at MIT," MIT Distinctive Collections, Cambridge, MA, March 9, 1970; Mildred Dresselhaus, interview, 1976, 207–210.

53. Mildred Dresselhaus, interview, 1976, 206–2010; MIT Institutional Research, "Women as a Percentage of Total Undergraduates, Graduate Students, and Faculty"; Mildred Dresselhaus, interview, 2007.

54. MIT Infinite History, "The Women of the Institute Panel Discussion"; Bix, *Girls Coming to Tech!*, 243–245; Robert M. Gray, "Coeducation at MIT: 1950s–1970s," unpublished manuscript, October 8, 2019, 41–42, https://ee.stanford.edu/~gray/Coeducation_MIT.pdf.

55. MIT Infinite History, "The Women of the Institute Panel Discussion"; Bix, *Girls Coming to Tech!*, 243–245.

56. Mildred Dresselhaus, interview, 2007; Mildred Dresselhaus, interview, 2004, 552; Diane Casselberry, "Energetic Woman Paves the Way for MIT Coeds," *Olean Times Herald*, April 19, 1972, 13; "Academic Honors," *Hartford Courant*, February 17, 1972, 2.

57. Mildred Dresselhaus, interview, 2007; US Department of Justice, "Equal Access to Education: Forty Years of Title IX," June 23, 2012, https://www.justice.gov/sites/default/files/crt/legacy/2012/06/20/titleixreport.pdf.

58. Mildred Dresselhaus, interview, 1976, 233; Bix, *Girls Coming to Tech!*, 244–249.

59. MIT Infinite History, "The Women of the Institute Panel Discussion."

60. Ad Hoc Committee on the Role of Women at MIT, "Role of Women Students at MIT," 1972, MIT Libraries Distinctive Collections, Cambridge, MA, 1–2; Bix, *Girls Coming to Tech!*, 245.

61. "Role of Women Students at MIT," 1–2.

62. "Role of Women Students at MIT," 1–3, 22–23.

63. Bix, *Girls Coming to Tech!*, 245.

64. "Role of Women Students at MIT," 57; Bix, *Girls Coming to Tech!*, 244; Kristen Sunter, "Mary Rowe Retiring from Role as Ombudsman," *Tech*, June 6, 2014, https://thetech.com/2014/06/06/ombudsman-v134-n27; Clarence Williams, *Technology and the Dream: Reflections on the Black Experience at MIT, 1941–1999* (Cambridge, MA: MIT Press, 2003), 3–4.

65. Diane Casselberry, "MIT Prof Opening Doors for Women," *Arizona Daily Star*, April 7, 1972, 55; Bix, *Girls Coming to Tech!*, 245.

66. Association of MIT Alumnae, "Millie: Trailblazer for Women of MIT."

67. Mildred Dresselhaus, interview, 1976, 271–272.

68. Reinaldo José Lopes, "Resistência física [Physical endurance]," *Folha de S. Paulo*, October 28, 2003, https://www1.folha.uol.com.br/folha/sinapse/ult1063u618.shtml.

69. Mildred Dresselhaus, interview, 1976, 273–299.

70. Paul Dresselhaus, "Growing Up with Millie," lecture, Celebrating Our Millie, MIT, Cambridge, MA, Nov. 26, 2017.

71. Marianne Dresselhaus Cooper, email to author, June 8, 2020.

72. Mildred Dresselhaus, interview, 1976, 273–281; Eliot Dresselhaus, interview with author, videoconference, April 29, 2020.

73. Mildred Dresselhaus, interview, 1976, 281–284; Paul Dresselhaus, email to author, May 26, 2020.

74. Paul Dresselhaus, email to author, Nov. 12, 2019.

75. Mildred Dresselhaus, interview by Steve Yalisove, Materials Research Society, "MRS Bulletin Interviews Mildred S. Dresselhaus-Graphite to Intercalation Compounds," September 25, 2013, https://youtu.be/F2eQcF9Dw3k.

CHAPTER 7

1. William Gibson, "First Cellular Phone Call Was Made 45 Years Ago," AARP, April 3, 2018, https://www.aarp.org/politics-society/history/info-2018/first-cell-phone-call.html; Sha Be Allah, "Today in Hip Hop History: Kool Herc's Party at 1520 Sedgwick Avenue 45 Years Ago Marks the Foundation of the Culture Known as Hip Hop," *The Source*, August 11, 2018, https://thesource.com/2018/08/11/today-in-hip-hop-history-kool-hercs-party-at-1520-sedgewick-avenue-45-years-ago-marks-the-foundation-of-the-culture-known-as-hip-hop.

2. Mildred Dresselhaus, interview by Shirlee Sherkow, 1976, transcript, Project on Women as Scientists and Engineers, MIT Libraries Distinctive Collections, Cambridge, MA, 242–243.

3. Mildred Dresselhaus, interview, 1976, 243–246; "Gender Gap," Dictionary.com, https://www.dictionary.com/browse/gender-gap; Sheila Widnall, "Millie's Impact on Women at MIT," prerecorded lecture, Celebrating Our Millie, MIT, Cambridge, MA, November 26, 2017.

4. Sheila Widnall, "Millie's Impact on Women at MIT."

5. Mildred Dresselhaus, interview by Clarence Williams, *Technology and the Dream: Reflections on the Black Experience at MIT, 1941–1999* (Cambridge, MA: MIT Press, 2003), 362–363.

6. Mildred Dresselhaus, interview, 1976, 251–252; MIT Office of Communications/Resource Development, "Endowed Professorships at MIT: A History," June 1984, Office of the Provost, MIT, Cambridge, MA, 194–195.

7. Mildred Dresselhaus, interview, 1976, 252.

8. Mildred Dresselhaus, interview by Brian Keegan, August 27, 2007, transcript, MIT Infinite History, Cambridge, MA, https://infinitehistory.mit.edu/video/mildred-s-dresselhaus.

9. Sheila Widnall, "Millie's Impact on Women at MIT."

10. Mildred Dresselhaus, interview by Steve Yalisove, Materials Research Society, "MRS Bulletin Interviews Mildred S. Dresselhaus—Graphite to Intercalation Compounds," September 25, 2013, https://youtu.be/F2eQcF9Dw3k.

11. Hiroshi Kamimura, "Graphite Intercalation Compounds," *Physics Today* 40, no. 12 (December 1987): 66–71, https://doi.org/10.1063/1.881095.

12. P. R. Wallace, "The Band Theory of Graphite," *Physical Review* 71, no. 622 (May 1, 1947): 622–634, https://doi.org/10.1103/PhysRev.71.622; Materials Research Society, "Mildred Dresselhaus—Nanocarbons from a Historical Perspective," YouTube video, February 14, 2017, https://youtu.be/xHeO9EYJHIs; Mildred Dresselhaus, interview by Magdolna Hargittai, *Candid Science IV: Conversations with Famous Physicists* (London: Imperial College Press, 2004), 565; N. B. Hannay et al., "Superconductivity in Graphitic Compounds," *Physical Review Letters* 14, no. 225 (February 15, 1965): 225–226, https://doi.org/10.1103/PhysRevLett.14.225.

13. Mildred Dresselhaus, interview, 2004, 565.

14. Alice Dragoon, "The 'What If?' Whiz," *MIT Technology Review*, April 23, 2013, https://www.technologyreview.com/s/513491/the-what-if-whiz/; D. D. L. Chung, "Mildred S. Dresselhaus (1930–2017)," *Nature* 534 (March 16, 2017): 316.

15. Mildred Dresselhaus, interview by Bernadette Bensaude-Vincent and Arne Hessenbruch, October 25, 2001, transcript, History of Recent

Science and Technology Project, Dibner Institute for the History of Science and Technology at MIT, Cambridge, MA, https://ethw.org/Oral-History:Mildred_Dresselhaus.

16. Mildred Dresselhaus, interview, 2004, 565; Deborah Chung, email to author, September 23, 2020.

17. Jesus de la Fuente, "CVD Graphene—Creating Graphene Via Chemical Vapour Deposition," Graphenea, https://www.graphenea.com/pages/cvd-graphene#.Xn7jrFApDOT.

18. Mildred S. Dresselhaus, "Modifying Materials by Intercalation," *Physics Today* 37, no. 3 (March 1, 1984): 63.

19. Mildred Dresselhaus, "Modifying Materials by Intercalation," 60.

20. Mildred Dresselhaus, interview, 2004, 565; Mildred Dresselhaus, interview, 2013.

21. Mildred Dresselhaus, "Modifying Materials by Intercalation," 62.

22. Mildred Dresselhaus, interview, 2007; Jim Handy, "How Big Is a Nanometer?" *Forbes*, December 14, 2011, https://www.forbes.com/sites/jimhandy/2011/12/14/how-big-is-a-nanometer/#53cee4826fb0; "Size of the Nanoscale," National Nanotechnology Initiative, https://www.nano.gov/nanotech-101/what/nano-size.

23. Hiroshi Kamimura, "Graphite Intercalation Compounds," 64–65.

24. Mildred Dresselhaus, interview, 2013.

25. Mildred Dresselhaus, interview, 2007; "Dr. Mildred S. Dresselhaus," National Academy of Engineering, https://www.nae.edu/Members Section/MemberDirectory/29468.aspx; "Women Elected to the National Academy of Engineering," Online Ethics Center for Engineering, Oct. 24, 2006, http://www.onlineethics.org/Topics/Diversity/DiverseResources/NAEwomen.aspx; "M. S. Dresselhaus," American Institute of Physics, https://history.aip.org/phn/11507001.html; "From NBS to NIST," National Institute of Standards and Technology, https://www.nist.gov/history/nist-100-foundations-progress/nbs-nist.

26. Shoshi Dresselhaus-Cooper, "Millie Dresselhaus Timeline," 2017.

27. Sam Merrill, "Women in Engineering," *Cosmopolitan*, April 1976, 162–166.

28. Merrill, "Women in Engineering," 164.

29. Mildred Dresselhaus, interview, 2007.

30. Joseph D. Martin, "Mildred Dresselhaus and Solid State Pedagogy at MIT," *Annalen der Physik* 2019, no. 531 (August 2019), https://doi.org/10.1002/andp.201900274.

31. Mildred Dresselhaus, interview, 2004, 560.

32. Natalie Angier, "Carbon Catalyst for Half a Century," *New York Times*, July 2, 2012, https://www.nytimes.com/2012/07/03/science/carbon-catalyst-for-half-a-century.html; Paul Dresselhaus, email to author, November 12, 2019.

33. Paul Dresselhaus, email, November 12, 2019.

34. Mildred Dresselhaus, interview, 2004, 555.

35. Paul Dresselhaus, email, November 12, 2019.

36. Eliot Dresselhaus, interview with author, videoconference, April 29, 2020.

37. Mildred Dresselhaus, interview, 2007.

38. Mildred Dresselhaus, interview, 2007.

39. Mildred Dresselhaus, interview, 2007; Denis Paiste, "Introducing the Materials Research Laboratory at MIT," *MIT News*, October 10, 2017, http://news.mit.edu/2017/introducing-mit-materials-research-laboratory-mrl-1010.

40. Mildred Dresselhaus, interview, 1976, 348; Mildred Dresselhaus, interview, 2004, 554–555.

41. Mildred Dresselhaus, interview, 2004, 555.

42. Mildred Dresselhaus, interview, 2004, 558.

43. Laura Doughty, interview by author, Wendell, MA, October 10, 2019.

44. Elizabeth Dresselhaus, email to author, November 15, 2019.

45. Mildred Dresselhaus, interview, 2004, 557.

46. "Mandelieu La Napoule," ProvenceWeb, https://www.provenceweb .fr/e/alpmarit/mandelie/mandelie.htm; Kamimura, "Graphite Intercalation Compounds," 65.

47. Mildred S. Dresselhaus and Gene Dresselhaus, "Intercalation Compounds of Graphite," *Advances in Physics* 30, no. 2 (1981): 139–326, https://doi.org/10.1080/00018738100101367; Mildred Dresselhaus, interview, 2004, 563.

48. Mildred Dresselhaus, interview, 2004, 566–567.

49. Mildred Dresselhaus, "Modifying Materials by Intercalation," 60–68; Mildred Dresselhaus, "Mildred Dresselhaus Biography," Kavli Prize, http://kavliprize.org/sites/default/files/%25nid%25/autobiagraphies _attachments/Mildred_Dresselhaus_Biography_0.pdf.

50. Dresselhaus, "Mildred Dresselhaus Biography;" Dresselhaus, "Modifying Materials by Intercalation," 63.

51. "Hydrogen Fuel Basics," US Department of Energy Office of Energy Efficiency and Renewable Energy, https://www.energy.gov/eere /fuelcells/hydrogen-fuel-basics; "Lithium-Ion Battery," Clean Energy Institute, University of Washington, https://www.cei.washington.edu /education/science-of-solar/battery-technology; "Graphene Supercapacitors: Introduction and News," Graphene-Info, https://www.graphene -info.com/graphene-supercapacitors.

52. Alice Dragoon, "The 'What If?' Whiz."

53. Mildred Dresselhaus, interview, 2013; Materials Research Society, "Nanocarbons from a Historical Perspective."

CHAPTER 8

1. Materials Research Society, "Mildred Dresselhaus—Nanocarbons from a Historical Perspective," YouTube video, February 14, 2017, https://youtu.be/xHeO9EYJHIs.

2. Deborah Chung, email to author, October 23, 2020; Pu-Woei Chen and Deborah D. L. Chung, "Carbon-Fiber-Reinforced Concrete Smart Structures Capable of Nondestructive Flaw Detection," *Proceedings SPIE, Smart Structures and Materials* 1916 (July 23, 1993): 22–30, https:// doi.org/10.1117/12.148502; Jeanie Chung, "Superconductor," *University of Chicago Magazine*, Summer 2015, https://mag.uchicago.edu /science-medicine/superconductor.

3. "Lamp Inventors 1880–1940: Carbon Filament Incandescent: Lighting a Revolution," Smithsonian National Museum of American History, https://americanhistory.si.edu/lighting/bios/swan.htm; Mildred S. Dresselhaus et al., *Graphite Fibers and Filaments* (Berlin: Springer-Verlag, 1988), 2; Morinobu Endo et al., "From Carbon Fibers to Nanotubes" in *Carbon Nanotubes: Preparation and Properties*, ed. Thomas W. Ebbesen (Boca Raton, FL: CRC Press, 1997), 39; Jonathan Martin, "Lewis H. Latimer 1848–1928," Contemporary Black Biography, Encyclopedia.com, https://www.encyclopedia.com/education/news-wires -white-papers-and-books/latimer-lewis-h-1848-1928; "February 10, 1874: Lewis Latimer Awarded First Patent," Mass Moments, https:// www.massmoments.org/moment-details/lewis-latimer-awarded-first -patent.html; "Lewis Latimer," National Inventors Hall of Fame, https:// www.invent.org/inductees/lewis-latimer.

4. Endo et al., "From Carbon to Nanotubes," 39–40; "High Performance Carbon Fibers," National Historic Chemical Landmarks, American Chemical Society, September 17, 2003, http://www.acs.org /content/acs/en/education/whatischemistry/landmarks/carbonfibers .html; Roger Bacon, "Growth, Structure, and Properties of Graphite Whiskers," *Journal of Applied Physics* 31, no. 2 (February 1960): 283–290, https://doi.org/10.1063/1.1735559.

5. Endo et al., "From Carbon Fibers to Nanotubes," 36–37; "What Is Carbon Fiber?" Zoltek; Mildred S. Dresselhaus et al., "Introduction to Carbon Materials" in *Carbon Nanotubes: Preparation and Properties*, ed. Thomas W. Ebbesen (Boca Raton, FL: CRC Press, 1997), 14; "How Is Carbon Fiber Made?" Carbon Fiber Education Center, Zoltek Corporation, https://zoltek.com/carbon-fiber/how-is-carbon-fiber-made; "High Performance Carbon Fibers," American Chemical Society; Chung, email, October 23, 2020.

6. Chung, email, October 23, 2020; D. D. L. Chung, "Comparison of Submicron-Diameter Carbon Filaments and Conventional Carbon Fibers as Fillers in Composite Materials," *Carbon* 39, no. 8 (2001): 1119–1125, https://doi.org/10.1016/S0008-6223(00)00314-6; Morinobu Endo et al., "Vapor-Grown Carbon Fibers (VGCFs): Basic Properties and Their Battery Applications," *Carbon* 39, no. 9 (2001): 1287–1297, https://doi.org/10.1016/S0008-6223(00)00295-5.

7. Yoong A. Kim et al., "Carbon Nanofibers," in *Springer Handbook of Nanomaterials*, ed. Robert Vajtai (Berlin: Springer-Verlag, 2013), 233–262; Chung, email, October 23, 2020.

8. Andrew Grant, "Mildred Dresselhaus (1930–2017)," *Physics Today*, February 23, 2017, https://physicstoday.scitation.org/do/10.1063/PT .5.9088/full/; Mildred Dresselhaus, interview by Magdolna Hargittai, *Candid Science IV: Conversations with Famous Physicists* (London: Imperial College Press, 2004), 566–567.

9. Andrew Grant, "Mildred Dresselhaus."

10. Mildred Dresselhaus et al., *Graphite Fibers and Filaments*, 1–4.

11. Mildred S. Dresselhaus and Gene Dresselhaus, "Intercalation Compounds of Graphite," *Advances in Physics* 30, no. 2 (1981): 139–326, https://doi.org/10.1080/00018738100101367; Mildred Dresselhaus et al., *Graphite Fibers and Filaments;* Mildred Dresselhaus, interview, 2004, 567.

12. Endo et al., "From Carbon to Nanotubes," 41; Mildred Dresselhaus, interview by Steve Yalisove, Materials Research Society, "MRS Bulletin Interviews Mildred S. Dresselhaus-Graphite to Intercalation Compounds," September 25, 2013, https://youtu.be/F2eQcF9Dw3k.

13. Mildred Dresselhaus, "Mildred Dresselhaus biography," Kavli Prize, http://kavliprize.org/sites/default/files/%25nid%25/autobiagraphies _attachments/Mildred_Dresselhaus_Biography_0.pdf.

14. "Organization and Governance," American Institute of Physics, https://www.aip.org/aip/leadership; "1982: American Institute of Physics Governing Board Meeting Minutes, 1931–1990," American Institute of Physics, https://www.aip.org/history-programs/niels-bohr

-library/collections/governing-board/1982; "1983: American Institute of Physics Governing Board Meeting Minutes, 1931–1990," American Institute of Physics, https://www.aip.org/history-programs/niels-bohr-library/collections/governing-board/1983; "1984: American Institute of Physics Governing Board Meeting Minutes, 1931–1990," American Institute of Physics, https://www.aip.org/history-programs/niels-bohr-library/collections/governing-board/1984.

15. Mildred S. Dresselhaus, "Perspectives on the Presidency of the American Physical Society," *Physics Today* 38, no. 7 (1985): 36–44, https://doi.org/10.1063/1.880980; "About APS," American Physical Society, https://www.aps.org/about/index.cfm; Maia Weinstock, "Chien-Shiung Wu: Courageous Hero of Physics" in *A Passion for Science: Stories of Discovery and Invention,* ed. Suw Charman-Anderson (London: Finding Ada, 2013), chap. 16, https://findingada.com/shop/a-passion-for-science-stories-of-discovery-and-invention.

16. Mildred Dresselhaus, "Perspectives," 37.

17. Mildred Dresselhaus, "Perspectives," 37; Mildred Dresselhaus, interview with Joseph D. Martin, transcript, American Institute of Physics, Niels Bohr Library and Archive, College Park, MD, June 24, 2014.

18. Laurie McNeil, "An Agent for Climate Change: Millie and Women in Science," lecture, Celebrating Our Millie, MIT, Cambridge, MA, November 26, 2017; "Site Visits," American Physical Society, https://www.aps.org/programs/women/sitevisits.

19. McNeil, "An Agent for Climate Change."

20. McNeil, "An Agent for Climate Change."

21. Mildred Dresselhaus, curriculum vitae, unpublished.

22. Marianne Dresselhaus Cooper, interview by author, Arlington, MA, April 27, 2018.

23. Paul Dresselhaus, email to author, November 12, 2019; Eliot Dresselhaus, interview by author, videoconference, April 29, 2020.

24. Melissa Clason, "The History of Disney World's Spaceship Earth," *WanderWisdom,* https://wanderwisdom.com/travel-destinations/The

-History-of-Spaceship-Earth; onstageDisney, "1982 Grand Opening of EPCOT Center," YouTube video, 23:17, June 30, 2018, https://youtu .be/wPRag-YygRE.

25. Tsz Yin Au et al., "TEA/AECOM 2018 Theme Index & Museum Index: Global Attractions Attendance Report," Teaconnect.org. Themed Entertainment Association, https://web.archive.org/web /20190523131129/http://www.teaconnect.org/images/files/328_572 762_190522.pdf; Kim Willis, "Disney World to Cut Theme Park Hours in September as Visits Drop amid COVID-19," *USA Today*, August 9, 2020, https://www.usatoday.com/story/travel/news/2020/08/09/disney -world-reduce-hours-september-visits-drop-amid-covid-19/333047 0001; Jennifer Fickley-Baker, "The Scientist Who Inspired the Name of Epcot's 'Spaceship Earth,'" Disney Parks Blog, September 28, 2012, https://disneyparks.disney.go.com/blog/2012/09/the-scientist-who -inspired-the-name-of-epcots-spaceship-earth.

26. "About Fuller: R. Buckminster Fuller, 1895–1983," Buckminster Fuller Institute, https://www.bfi.org/about-fuller/biography; Lauren Beale, "It's a Brave New World for the Former Aldous Huxley Estate," *South Florida Sun Sentinel*, June 11, 2018, https://www.sun-sentinel .com/real-estate/prime-property/sfl-it-s-a-brave-new-world-for-the -former-aldous-huxley-estate-20180613-story.html; Adam Rome, "The Launch of Spaceship Earth," *Nature*, 527 (November 26, 2015): 443– 445.

27. "About Fuller: Geodesic Domes," Buckminster Fuller Institute, https://www.bfi.org/about-fuller/big-ideas/geodesic-domes; Eric W. Weisstein, "Geodesic Dome," MathWorld, https://mathworld.wolfram .com/GeodesicDome.html; "Spaceship Earth," Walt Disney World, https://disneyworld.disney.go.com/attractions/epcot/spaceship-earth.

28. Jonathan Glancey, "The Story of Buckminster Fuller's Radical Geodesic Dome," BBC, October 4, 2014, http://www.bbc.com/culture /story/20140613-spaceship-earth-a-game-of-domes; Eric W. Weisstein, "Truncated Icosahedron," MathWorld, https://mathworld.wolfram .com/TruncatedIcosahedron.html; Mildred S. Dresselhaus, Gene Dres-selhaus, and Paul C. Eklund, *Science of Fullerenes and Carbon Nano-tubes* (San Diego: Academic Press, 1996), 7; H. W. Kroto et al., "C_{60}:

Buckminsterfullerene," *Nature* 318 (November 14, 1985): 162–163, https://doi.org/10.1038/318162a0.

29. Richard E. Smalley, "Discovering the Fullerines," Nobel Lecture, Stockholm, Sweden, December 7, 1996, https://www.nobelprize.org /uploads/2018/06/smalley-lecture.pdf.

30. George W. Hart, "Archimedean Polyhedra," Virtual Polyhedra, 1996, https://www.georgehart.com/virtual-polyhedra/archimedean-info .html; George W. Hart, "Piero della Francesca's Polyhedra," Virtual Polyhedra, 1998, https://www.georgehart.com/virtual-polyhedra/piero .html; M. S. Dresselhaus et al., *Science of Fullerenes*, 2–3; Smalley, "Discovering the Fullerines."

31. Mildred Dresselhaus, "Mildred Dresselhaus Biography"; Mildred Dresselhaus, interview by the Kavli Foundation, "2012 Kavli Prize in Nanoscience: A Discussion with Mildred Dresselhaus," August 2012, https://www.kavlifoundation.org/science-spotlights/kavli-prize-2012 -dresselhaus#.XjEWVlBOnOR; Mildred Dresselhaus, interview, 2004, 562–563; T. Venkatesan et al., "Measurement of Thermodynamic Parameters of Graphite by Pulsed-Laser Melting and Ion Channeling," *Physical Review Letters* 53, no. 4 (July 23, 1984): 360–363.

32. Mildred Dresselhaus, interview, 2004, 562–563.

33. Materials Research Society, "Nanocarbons from a Historical Perspective."

34. Materials Research Society, "Nanocarbons from a Historical Perspective."

35. Eric A. Rohlfing, D. M. Cox, and A. Kaldor, "Production and Characterization of Supersonic Carbon Cluster Beams," *Journal of Chemical Physics* 81 (October 1, 1984): 3322–3330, https://doi.org/10.1063 /1.447994; Smalley, "Discovering the Fullerenes."

36. Mildred Dresselhaus, interview, 2004, 563.

37. Smalley, "Discovering the Fullerenes."

38. National Research Council, "Ion Implantation and Surface Modification," in *Plasma Processing and Processing Science* (Washington, DC:

National Academies Press, 1995), 15–18. https://doi.org/10.17226
/9854; Dresselhaus, interview, 2004, 562–563; Materials Research
Society, "Nanocarbons from a Historical Perspective"; Mildred Dresselhaus, "Mildred Dresselhaus biography."

39. Mildred Dresselhaus, interview, August 2012.

40. "Discovery of Fullerenes," National Historic Chemical Landmarks,
American Chemical Society, October 11, 2010, http://www.acs.org
/content/acs/en/education/whatischemistry/landmarks/fullerenes
.html; Hugh Aldersey-Williams, *The Most Beautiful Molecule: The Discovery of the Buckyball* (New York: Wiley, 1995), 52–90.

41. Harold W. Kroto, "Symmetry, Space, Stars, and C_{60}," Nobel Lecture, Stockholm, Sweden, December 7, 1996, https://www.nobelprize
.org/uploads/2018/06/kroto-lecture.pdf.

42. ACS, "Discovery of Fullerenes"; W. Krätschmer et al., "Solid C60:
a new form of carbon." *Nature* 347 (Sept. 27, 1990): 354–358, https://
doi.org/10.1038/347354a0; David R. M. Walton and Harold W. Kroto,
"Fullerene," *Encyclopedia Britannica*, https://www.britannica.com/science
/fullerene.

43. "Ado Jorio," ResearchGate, https://www.researchgate.net/profile
/Ado_Jorio; Ado Jorio, "A Journey with the Queen of Carbon," Celebrating Millie, PubPub, March 20, 2018, https://millie.pubpub.org
/pub/lboloi9l. This is an edited version of the speech Jorio gave at
Celebrating Our Millie, the November 26, 2017, event at MIT, Cambridge, MA.

44. Barnaby J. Feder, "The Nobel Prize that Wasn't," *New York Times*,
September 18, 2007, https://bits.blogs.nytimes.com/2007/09/18/the
-nobel-prize-that-wasnt; Markian Hawryluk, "Discovery of Buckyballs
a Nobel Effort by Professors," *Houston Chronicle*, May 23, 2016, https://
www.chron.com/local/history/medical-science/article/Discovery-of
-Buckyballs-a-Nobel-effort-by-7939221.php.

45. Barnaby J. Feder, "Richard E. Smalley, 62, Dies; Chemistry Nobel
Winner," *New York Times*, October 29, 2005, https://www.nytimes.com
/2005/10/29/science/richard-e-smalley-62-dies-chemistry-nobel-winner

.html; Richard Feynman, "There's Plenty of Room at the Bottom," *Caltech Engineering and Science* 23, no. 5 (February 1960): 22–36, http:// calteches.library.caltech.edu/1976/1/1960Bottom.pdf.

46. Kenneth Chang, "A Prodigious Molecule and Its Growing Pains," *New York Times*, October 10, 2000, https://www.nytimes.com /2000/10/10/science/a-prodigious-molecule-and-its-growing-pains .html.

47. Mildred Dresselhaus, curriculum vitae, unpublished.

48. Killian Faculty Achievement Award Committee (Bruno Coppi, David II. Marks, Bruce Mazlish, Edward B. Roberts, William L. Porter) to MIT Faculty, Recommendation of Professor Mildred S. Dresselhaus, May 21, 1986, MIT Institute Events, https://killianlectures.mit.edu /sites/default/files/images/Killian%20Citation%20M%20Dresselhaus .pdf.

49. Mildred Dresselhaus, curriculum vitae, unpublished; Mildred S. Dresselhaus et al., *Graphite Fibers and Filaments* (Berlin: Springer-Verlag, 1988).

50. Laura Doughty, interview by author, Wendell, MA, October 10, 2019.

51. Margaret W. Rossiter, *Women Scientists in America: Forging a New World since 1972* (Baltimore: Johns Hopkins University Press, 2012), 109–110, 133.

52. Mildred Dresselhaus, curriculum vitae, unpublished; National Research Council, "Women in Science and Engineering: Increasing Their Numbers in the 1990s: A Statement on Policy and Strategy" (Washington, DC: National Academies Press, 1991), https://doi.org /10.17226/1878.

53. Michelle Buchanan, "Millie Serving Society," lecture, Celebrating Our Millie, MIT, Cambridge, MA, Nov. 26, 2017.

54. "History," National High Magnetic Field Laboratory, https:// nationalmaglab.org/about/history.

55. Alice Dragoon, "The 'What If?' Whiz," *MIT Technology Review*, April 23, 2013, https://www.technologyreview.com/s/513491/the-what -if-whiz/.

56. Dragoon, "The 'What If?' Whiz."

57. Dragoon, "The 'What If?' Whiz."

58. David L. Chandler, "Explained: Phonons," *MIT News*, July 8, 2010, http://news.mit.edu/2010/explained-phonons-0706; Mildred Dressel- haus, "Mildred Dresselhaus Biography."

59. "The President's National Medal of Science: Recipient Details— Mildred S. Dresselhaus," National Science Foundation, https://www .nsf.gov/od/nms/recip_details.jsp?recip_id=110; Associated Press, "President Honors 30 for Research," *New York Times*, November 14, 1990, A22. In addition to the twenty National Medal of Science recipi- ents, ten individuals earned the National Medal of Technology.

60. NSF, "Medal of Science Details—Dresselhaus."

61. "The President's National Medal of Science: Recipients," National Science Foundation, https://www.nsf.gov/od/nms/recipients.jsp.

62. Mildred Dresselhaus, "Mildred Dresselhaus Biography."

63. Mildred Dresselhaus, "Mildred Dresselhaus Biography"; Jayeeta Lahiri, "The Queen of Carbon! Mildred Dresselhaus (1930–2017)," *Resonance—Journal of Science Education* 24, no. 3 (March 2019): 263– 272, https://www.ias.ac.in/article/fulltext/reso/024/03/0263-0272.

64. Ting Guo et al., "Self-Assembly of Tubular Fullerenes," *Journal of Physical Chemistry* 99, no. 27 (July 1, 1995): 10694–10697, https:// doi.org/10.1021/j100027a002; Sumio Iijima, "Helical Microtubules of Graphitic Carbon," *Nature* 354 (November 7, 1991): 56–58, https:// doi.org/10.1038/354056a0; Mildred S. Dresselhaus, Gene Dresselhaus, and Riichiro Saito, "C_{60}-Related Tubules," *Solid State Communications* 84, no. 1–2 (October 1992): 201–205, https://doi.org/10.1016/0038 -1098(92)90325-4; Mildred S. Dresselhaus, Gene Dresselhaus, and Riichiro Saito, "Carbon Fibers Based on C_{60} and Their Symmetry," *Physical Review B* 45, no. 11 (March 15, 1992): 6234–6242, https://doi

.org/10.1103/PhysRevB.45.6234; Ivan Amato, "The Soot That Could Change the World," *Fortune*, June 25, 2001, https://archive.fortune .com/magazines/fortune/fortune_archive/2001/06/25/305482/index .htm.

65. Mildred Dresselhaus, "Mildred Dresselhaus Biography"; Sumio Iijima, "Weizmann Institute Memorial Lecture for Millie Dresselhaus," Celebrating Millie, PubPub, June 21, 2018, https://millie.pubpub.org /pub/bqe71oiz; Mildred Dresselhaus, interview by Brian Keegan, August 27, 2007, transcript, MIT Infinite History, Cambridge, MA, https://infinitehistory.mit.edu/video/mildred-s-dresselhaus.

66. Michael F. L. De Volder et al., "Carbon Nanotubes: Present and Future Commercial Applications," *Science* 339, no. 535 (February 1, 2013): 535–539, http://doi.org/10.1126/science.1222453; Cheap Tubes, Inc., "Applications of Carbon Nanotubes," AZO Nano, April 23, 2018, https://www.azonano.com/article.aspx?ArticleID=4842.

67. "Carbon Nanotubes (CNT) Market by Type, Method, Application— Global Forecast to 2023," Research and Markets, October 2018, https://www.researchandmarkets.com/research/jrjnq4/the_global _carbon?w=5.

68. De Volder et al., "Carbon Nanotubes"; Soehil Jafari, "Engineering Applications of Carbon Nanotubes," in *Carbon Nanotube Reinforced Polymers: From Nanoscale to Macroscale*, ed. Roham Rafiee (Amsterdam: Elsevier, 2018), 25–40.

69. Marc Monthioux and V. L. Kuznetsov, "Who Should Be Given the Credit for the Discovery of Carbon Nanotubes?" *Carbon* 44, no. 9 (August 2006): 1621–1623, http://doi.org/10.1016/j.carbon.2006 .03.019; Endo et al., "From Carbon Fibers to Nanotubes," 40–41.

70. Monthioux and Kuznetsov, "Credit for Carbon Nanotubes?" 1621–1623; Agnès Oberlin, Morinobu Endo, and Tsuneo Koyama, "Filamentous Growth of Carbon through Benzene Decomposition," *Journal of Crystal Growth* 32, no. 3 (March 1976): 335–349, https:// doi.org/10.1016/0022-0248(76)90115-9; Nicole Grobert, "Carbon Nanotubes—Becoming Clean," *Materials Today* 10, no. 1–2 (January–

February 2007): 28–35, https://doi.org/10.1016/S1369-7021(06)71789 -8; Mildred Dresselhaus, interview by Paul S. Weiss, "A Conversation with Prof. Mildred Dresselhaus: A Career in Carbon Nanomaterials," *ACS Nano* 3, no. 9 (September 2009): 2435; Mildred Dresselhaus, "Graphene: A Journey through Carbon Nanoscience," *MRS Bulletin* 37, no. 12 (December 2012): 1319, https://doi-org.libproxy.mit.edu/10.1557 /mrs.2012.301.

71. Iijima, "Helical Microtubules of Graphitic Carbon"; Monthioux and Kuznetsov, "Credit for Carbon Nanotubes?" 1621–1623; Materials Research Society, "Nanocarbons from a Historical Perspective."

72. Mildred Dresselhaus, Gene Dresselhaus, and Riichiro Saito, "Carbon Fibers Based on C_{60} and Their Symmetry," *Physical Review B* 45, no. 11 (March 1992): 6234–6242; Riichiro Saito et al., "Electronic Structure of Carbon Fibers Based on C_{60}," *MRS Proceedings* 247 (1992): 333. Riichiro Saito et al., "Electronic Structure of Chiral Graphene Tubules," *Applied Physics Letters* 60, no. 18 (May 1992): 2204–2206.

73. Riichiro Saito, "Early Times in Carbon Nanotubes," lecture, Celebrating Our Millie, MIT, Cambridge, MA, November 26, 2017; "Riichiro Saito," Tohoku University Department of Physics, https://flex .phys.tohoku.ac.jp/~rsaito/rsaito-e.html.

74. Saito, "Early Times in Carbon Nanotubes."

75. Mildred Dresselhaus, interview, August 2012; Saito et al., "Electronic Structure of Carbon Fibers Based on C_{60}"; Saito et al., "Electronic Structure of Chiral Graphene Tubules."

76. Earl Boysen and Nancy Boysen, *Nanotechnology for Dummies*, 2nd ed. (Hoboken, NJ: Wiley, 2011), 40–42.

77. Saito et al., "Electronic Structure of Chiral Graphene Tubules"; Boysen and Boysen, *Nanotechnology for Dummies*, 40–42.

78. Reshef Tenne, "Weizmann Institute Memorial Lecture for Millie Dresselhaus," Celebrating Millie, PubPub, June 21, 2018, https:// millie.pubpub.org/pub/bqe71oiz.

79. Mildred Dresselhaus, interview, August 2007.

80. Sumio Iijima and Toshinari Ichihashi, "Single-Shell Carbon Nanotubes of 1-nm Diameter," *Nature* 363 (June 17, 1993): 603–605, https://doi.org/10.1038/363603a0; Donald Bethune et al., "Cobalt-Catalysed Growth of Carbon Nanotubes with Single-Atomic-Layer Walls," *Nature* 363 (June 17, 1993): 605–607, https://doi.org/10.1038/363605a0; "Donald S. Bethune," IBM Research, https://researcher.watson.ibm.com /researcher/view.php?person=us-dbethune; Apparo M. Rao et al., "Diameter-Selective Raman Scattering from Vibrational Modes in Carbon Nanotubes," *Science* 275, no. 5297 (January 10, 1997): 187–191, http://doi.org/10.1126/science.275.5297.187.

81. M. Dresselhaus, G. Dresselhaus, and Eklund, *Science of Fullerenes and Carbon Nanotubes*; "Obituary of Peter Clay Eklund, 64," StateCollege. com, August 21, 2009, http://www.statecollege.com/obituary/detail /obituary-of-peter-clay-eklund--64,155.

82. Mildred Dresselhaus, "Professor Dresselhaus' Closing Remarks," lecture, 19th Science in Japan Forum, Japan Society for the Promotion of Science, Cosmos Club, Washington, DC, October 3, 2014, https:// jspsusa.org/wp/wp-content/uploads/2014/03/19SiJFRemarks.pdf.

CHAPTER 9

1. Marcie Black, email to author, July 26, 2019; Marcie Black, email to author, May 6, 2020; "What Is the Kyoto Protocol?" United Nations Framework Convention on Climate Change secretariat, https:// unfccc.int/kyoto_protocol.

2. Black, email, May 6, 2020.

3. Marcie Black, "Dr. Millie Dresselhaus: One in Ten Million Scientist, Amazing Violin Player, But So Much More Than That," poster, Celebrating Our Millie, MIT, Cambridge, MA, November 26, 2017, https:// millie.pubpub.org/pub/4msu13pj/release/1.

4. Hui-Ming Cheng, Quan-Hong Yang, and Chang Liu, "Hydrogen Storage in Carbon Nanotubes," *Carbon* 39 no. 10 (August 2001): 1447–1454, https://doi.org/10.1016/S0008-6223(00)00306-7; Chang Liu et al., "Hydrogen Storage in Single-Walled Carbon Nanotubes at Room

Temperature," *Science* 286, no. 5442 (1999): 1127–1129, http://doi
.org/10.1126/science.286.5442.1127.

5. NanoTube—The National Nanotechnology Initiative, "Changing
the World with Nano-Textured Silicon: A Conversation with Dr. Marcie Black," YouTube video, March 11, 2019, https://youtu.be/1xG9wO
-5Ksg.

6. Black, "Dr. Millie Dresselhaus."

7. Black, email, May 6, 2020; Black, "Dr. Millie Dresselhaus."

8. Black, email, July 26, 2019.

9. Black, email, July 26, 2019.

10. Black, email, July 26, 2019.

11. "Institute Professor Emerita Mildred Dresselhaus, a Pioneer in the
Electronic Properties of Materials, Dies at 86," *MIT News*, February 21,
2017, http://news.mit.edu/2017/institute-professor-emerita-mildred
-dresselhaus-dies-86-0221.

12. Eliot Dresselhaus, interview by author, videoconference, April 29,
2020.

13. Andrew Grant, "Mildred Dresselhaus (1930–2017)," *Physics Today*,
February 23, 2017, http://doi.org/DOI:10.1063/PT.5.9088.

14. Mildred Dresselhaus, interview by Clarence Williams, *Technology
and the Dream: Reflections on the Black Experience at MIT, 1941–1999*
(Cambridge, MA: MIT Press, 2003), 359.

15. Amanda Schaffer, "The Remarkable Career of Shirley Ann Jackson," *MIT Technology Review*, December 19, 2017, https://www
.technologyreview.com/2017/12/19/146775/the-remarkable-career-of
-shirley-ann-jackson.

16. Shirley Ann Jackson, prerecorded lecture, Celebrating Our Millie,
MIT, Cambridge, MA, Nov. 26, 2017.

17. Jackson, lecture.

18. Mildred Dresselhaus, interview, 2003, 359.

19. Jackson, lecture.

20. "Brief History of Thermoelectrics," Northwestern University Materials Science and Engineering, http://thermoelectrics.matsci .northwestern.edu/thermoelectrics/history.html; David L. Chandler, "Explained: Thermoelectricity," *MIT News*, April 27, 2010, http:// news.mit.edu/2010/explained-thermoelectricity-0427.

21. Jospeh P. Heremans et al., "When Thermoelectrics Reached the Nanoscale," *Nature Nanotechnology* 8 (July 2013): 471–473.

22. John G. Stockholm, "A Call from the Navy: Millie and Thermoelectrics," lecture, Celebrating Our Millie, MIT, Cambridge, MA, November 26, 2017.

23. Chandler, "Explained: Thermoelectricity"; Gang Chen, email to author, June 3, 2020.

24. Alice Dragoon, "The 'What If?' Whiz," *MIT Technology Review*, April 23, 2013, https://www.technologyreview.com/s/513491/the-what -if-whiz/.

25. Mildred Dresselhaus, interview by Brian Keegan, August 27, 2007, transcript, MIT Infinite History, Cambridge, MA, https://infinitehistory .mit.edu/video/mildred-s-dresselhaus.

26. "Mildred Dresselhaus Biography," Kavli Prize, http://kavliprize.org /sites/default/files/%25nid%25/autobiagraphies_attachments/Mildred _Dresselhaus_Biography_0.pdf; Lyndon D. Hicks and Mildred S. Dresselhaus, "Effect of Quantum-Well Structures on the Thermoelectric Figure of Merit," *Physical Review B* 47, no. 19 (May 15, 1993): 12727–12731; Lyndon D. Hicks and Mildred S. Dresselhaus, "Thermoelectric Figure of Merit of a One-Dimensional Conductor," *Physical Review B* 47, no. 24 (May 15, 1993): 16631–16634; Lyndon D. Hicks et al., "Use of Quantum-Well Superlattices to Obtain a High Figure of Merit from Nonconventional Thermoelectric Materials," *Applied Physics Letters* 63, no. 23 (December 6, 1993): 3230–3232; David L. Chandler, "Thermoelectric Materials Are One Key to Energy Savings," *MIT Tech Talk*, MIT News Office, November 20, 2007, http://news.mit.edu/2007 /nanoenergy-1120; Chen, email, June 3, 2020.

27. Oded Rabin, "Thermoelectrics Research in MGM," lecture, Celebrating Our Millie, MIT, Cambridge, MA, November 26, 2017; Chen, email, June 3, 2020.

28. Heremans et al., "Thermoelectrics," 473.

29. Jennifer Chu, "Turning Up the Heat on Thermoelectrics," *MIT News*, May 25, 2018, http://news.mit.edu/2018/materials-heated -magnetic-fields-thermoelectrics-0525.

30. Mildred Dresselhaus, curriculum vitae.

31. Gang Chen, interview by author, Cambridge, MA, August 13, 2019.

32. U.S. Patent numbers US8168879B2, US8865995B2, US8293168B2, US7586033B2, US7255846B2, Google Patents.

33. Chen, interview, August 13, 2019.

34. Chen, interview, August 13, 2019.

35. Gang Chen, letter to the editor, *MIT Technology Review*, June 28, 2013, https://www.technologyreview.com/2013/06/18/177742/letters -34.

36. Mildred Dresselhaus, interview by Magdolna Hargittai, *Candid Science IV: Conversations with Famous Physicists* (London: Imperial College Press, 2004), 569.

37. Mildred Dresselhaus, interview by Martha A. Cotter and Mary S. Hartman, *Talking Leadership: Conversations with Powerful Women* (New Brunswick, NJ: Rutgers University Press, 1999), 68; "Value of $5 from 1951 to 2020, Inflation Calculator," Official Inflation Data, Alioth, https://www.officialdata.org/us/inflation/1951?amount=5.

38. Mildred Dresselhaus, interview, 1999, 68.

39. Mildred S. Dresselhaus, "The AAAS Celebrates Its 150th," *Science* 282, no. 5397 (December 18, 1998): 2186–2190, http://doi.org /10.1126/science.282.5397.2186.

40. "Dresselhaus Speaks of Goals during her AAAS Term," *MIT Tech Talk*, MIT News Office, October 23, 1996, http://news.mit.edu/1996

/dresselhaus-1023; Mildred Dresselhaus, interview by Harry Kroto, Vega Science Trust, 2001, http://www.vega.org.uk/video/programme/20; David Malakoff, "Clinton's Science Legacy: Ending on a High Note," *Science* 290, no. 5500 (December 22, 2000): 2236–2236, http://doi.org/10.1126/science.290.5500.2234.

41. William J. Clinton, "Remarks by the President for the American Association for the Advancement of Science," lecture, American Association for the Advancement of Science annual meeting, February 13, 1998, https://clintonwhitehouse4.archives.gov/textonly/WH/New/html/19980213-26754.html

42. Mihail C. Roco, "The Long View of Nanotechnology Development: The National Nanotechnology Initiative at 10 Years," *Journal of Nanoparticle Research* 13, no. 2 (February 2011): 427–445, http://doi.org/10.1007/s11051-010-0192-z.

43. Irwin Goodwin, "As Term Nears End, Clinton Names Dresselhaus to Strengthen Support for DOE Science," *Physics Today* 53, no. 6 (June 1, 2000): 48–49, https://doi.org/10.1063/1.1306368.

44. Deborah Halber, "Dresselhaus Sworn In as Head of DOE's Office of Science," *MIT News*, October 23, 2000, http://news.mit.edu/2000/dresselhaus-sworn-head-does-office-science.

45. Halber, "Dresselhaus Sworn In"; Mildred Dresselhaus, interview, 2004, 553–554.

46. Judy Jackson, "Millie Comes to Fermilab," *FermiNews* 23, no. 18 (October 20, 2000), https://www.fnal.gov/pub/ferminews/ferminews00-10-20/p1.html.

47. Laura Doughty, interview by author, Wendell, MA, October 10, 2019; Mildred Dresselhaus, curriculum vitae, unpublished; Shoshi Dresselhaus-Cooper and Leora Dresselhaus-Marais, via Marianne Dresselhaus Cooper, email to author, June 8, 2020.

48. Shoshi Dresselhaus-Cooper, "Millie Dresselhaus Timeline," 2017; Shoshi Dresselhaus-Cooper and Leora Dresselhaus-Marais, via Marianne Dresselhaus Cooper, email, June 8, 2020.

49. Shoshi Dresselhaus-Cooper, email to author, August 18, 2019; Shoshi Dresselhaus-Cooper, email to author, August 22, 2019; Mildred Dresselhaus, interview, August 2007.

50. Shoshi Dresselhaus-Cooper, email, August 18, 2019; Shoshi Dresselhaus-Cooper, email, August 22, 2019.

51. Shoshi Dresselhaus-Cooper, email, August 18, 2019; Shoshi Dresselhaus-Cooper, email, August 22, 2019; Shoshi Dresselhaus-Cooper and Leora Dresselhaus-Marais via Marianne Dresselhaus Cooper, email, June 8, 2020.

52. Mildred Dresselhaus, curriculum vitae; "AIP Board Chairs," American Institute of Physics, https://www.aip.org/aip/board-chairs; Mildred Dresselhaus, "Basic Research Needs for the Hydrogen Economy," lecture at the American Physical Society meeting, Montreal, Canada, March 23, 2004, https://www.aps.org/meetings/multimedia/upload /Mildred_Dresselhaus.pdf.

53. Andre K. Geim, "Graphene Prehistory," *Physica Scripta* 2012, no. T146 (January 31, 2012): 014003, https://doi.org/10.1088/0031-8949 /2012/T146/014003; Phillip R. Wallace, "The Band Theory of Graphite," *Physical Review* 71, no. 9 (May 1, 1947): 622–634; Mildred S. Dresselhaus and Paulo T. Araujo, "Perspectives on the 2010 Nobel Prize in Physics for Graphene," *ACS Nano* 4, no. 11 (November 23, 2010): 6297, https://doi.org/10.1021/nn1029789.

54. Mildred Dresselhaus and Paulo Araujo, "Perspectives," 6298.

55. Mildred S. Dresselhaus and Paulo T. Araujo, "Perspectives."

56. Hanns-Peter Boehm et al., "Dünnste Kohlenstoff-Folien," *Zeitschrift für Naturforschung B*. 17, no. 3 (March, 1962): 150–153, https:// doi.org/10.1515/znb-1962-0302; Hanns-Peter Boehm et al., "Das Adsorptionsverhalten sehr dünner Kohlenstoff-Folien," *Journal of Inorganic and General Chemistry* 316, no. 3–4 (July 1962): 119–127, https:// doi.org/10.1002/zaac.19623160303; Mildred Dresselhaus and Paulo Araujo, "Perspectives," 6297–6300; Geim, "Graphene Prehistory"; Andre Geim and Konstantin Novoselov, "The rise of graphene." *Nature Materials* 6 (March 2007): 183–191, https://doi.org/10.1038/nmat1849.

57. David L. Chandler, "A Material for All Seasons," *MIT Tech Talk* 53, no. 54 (May 6, 2009): 1, http://news.mit.edu//2009/techtalk53-24.pdf.

58. Mildred Dresselhaus and Paulo Araujo, "Perspectives," 6297–6298; Geim, "Graphene Prehistory."

59. Mildred Dresselhaus and Paulo Araujo, "Perspectives," 6297–6299; Giles Whittell, "The Godfather of Graphene," *Economist* (September/October 2014), https://www.1843magazine.com/content/features/giles-whittell/andre-geim; Konstantin S. Novoselov et al., "Electric Field Effect in Atomically Thin Carbon Films," *Science* 306, no. 5696 (October 22, 2004): 666–669, http://doi.org/10.1126/science.1102896.

60. Geim and Novoselov, "Rise of Graphene," 184; Kenneth Chang, "Thin Carbon Is In: Graphene Steals Nanotubes' Allure," *New York Times*, April 10, 2017, https://www.nytimes.com/2007/04/10/science/10grap.html.

61. Andre Geim, "Random Walk to Graphene," Nobel lecture, Stockholm, Sweden, December 8, 2010, NobelPrize.org, https://www.nobelprize.org/prizes/physics/2010/geim/lecture; Materials Research Society, "Mildred Dresselhaus—Nanocarbons from a Historical Perspective," YouTube video, February 14, 2017, https://youtu.be/xHeO9EYJHIs.

62. John Colapinto, "Material Question," *New Yorker*, December 15, 2014, https://www.newyorker.com/magazine/2014/12/22/material-question.

63. Chandler, "A Material for All Seasons," 4.

64. Jennifer Chu, "Insulator or Superconductor? Physicists Find Graphene Is Both," *MIT News*, March 5, 2018, http://news.mit.edu/2018/graphene-insulator-superconductor-0305; Earl Boysen and Nancy Boysen, *Nanotechnology for Dummies*, 2nd ed. (Hoboken, NJ: Wiley, 2011), 43–44; Jing Kong, email to author, May 29, 2020; Angela Chen, "Behind the Hype: Experts Explain the Science behind Graphene, the New Supermaterial," *Verge*, January 24, 2018, https://www.theverge.com/2018/1/24/16927224/graphene-materials-les-johnson-joseph-meany-book; Jennifer Chu, "Researchers 'Iron Out' Graphene's

Wrinkles," *MIT News*, April 3, 2017, http://news.mit.edu/2017/iron-out
-graphene-wrinkles-conductive-wafers-0403.

65. "Dresselhaus Wins L'OREAL-UNESCO for Women in Science Prize,"
SEED, February 22, 2007, Wayback Machine, https://web.archive
.org/web/20160702063306/http://seedmagazine.com/content/article
/dresselhaus_wins_loreal-unesco_for_women_in_science_prize.

66. Mildred Dresselhaus, curriculum vitae.

67. Mildred Dresselhaus, interview by Paul S. Weiss, "A Conversation
with Prof. Mildred Dresselhaus: A Career in Carbon Nanomaterials,"
ACS Nano 3, no. 9 (September 2009), 2436.

68. Jing Kong, interview with author, Cambridge, MA, September 10,
2019.

69. Kong, interview, 2019; Mildred Dresselhaus, curriculum vitae.

70. Kong, interview, 2019.

71. Mildred Dresselhaus, curriculum vitae; Kong, interview, 2019.

72. Kong, interview, 2019; Stephen Salk, MIT Human Resources,
email to author, May 28, 2020.

73. Kong, interview, 2019.

74. Kong, interview, 2019.

75. "The Nobel Prize in Physics 2010," NobelPrize.org, Nobel
Media AB 2020, https://www.nobelprize.org/prizes/physics/2010/prize
-announcement.

76. Mildred Dresselhaus and Paulo Araujo, "Perspectives," 6301.

77. Andre Geim, "Walk to Graphene"; Konstantin S. Novoselov, "Graphene; Materials in the Flatland," Nobel lecture, Stockholm, Sweden,
December 8, 2010, NobelPrize.org, https://www.nobelprize.org/prizes
/physics/2010/novoselov/lecture; Laura Doughty, phone interview by
author, November 28, 2020.

78. Doughty, interview, October 10, 2019.

79. Mildred Dresselhaus, interview, 2004, 563.

CHAPTER 10

1. MIT Department of Electrical Engineering and Computer Science, "Honoring Millie," *MIT News*, December 8, 2010, http://news.mit.edu /2010/dresselhaus-birthday.

2. Irene Yong Rong Huang, MIT Department of Electrical Engineering and Computer Science, email to author, May 19, 2020; MIT EECS, "Honoring Millie."

3. Keith O'Brien, "Pioneering Woman Physicist, Cited for Her Research, Mentoring," *Boston Globe*, March 5, 2007, 19; Mildred Dresselhaus, interview by Jenni Murray, "The Age of Reason," BBC, December 29, 2012, https://www.bbc.co.uk/programmes/p012bp6b.

4. MIT EECS, "Honoring Millie."

5. O'Brien, "Pioneering Woman Physicist," 19.

6. Asma Khalid, "Visionaries: MIT's Alan Guth Made a 'Spectacular Realization' about the Universe," WBUR, February 26, 2015, https:// www.wbur.org/news/2015/02/26/visionaries-alan-guth-mit.

7. Read Schusky, interview by author, Cambridge, MA, November 14, 2019; Laura Doughty, email to author, May 26, 2020.

8. Laura Doughty, interview by author, Wendell, MA, October 10, 2019.

9. Mark Anderson, "The Queen of Carbon," *IEEE Spectrum*, April 28, 2015, https://spectrum.ieee.org/geek-life/profiles/mildred-dresselhaus -the-queen-of-carbon; Schusky, interview, November 14, 2019; Doughty, interview, October 10, 2019.

10. Doughty, interview, October 10, 2019.

11. Gang Chen, interview by author, Cambridge, MA, August 13, 2019; Marcie Black, email to author, July 26, 2019;

12. Schusky, interview, November 14, 2019.

13. "Dresselhaus Speaks of Goals during Her AAAS term," *MIT Tech Talk*, MIT News Office, October 23, 1996, http://news.mit.edu/1996 /dresselhaus-1023; Doughty, interview, October 10, 2019.

14. Mildred Dresselhaus, "Memories of Arthur von Hippel, 1898–2003," *MRS Bulletin* 39 (November 2014): 1000; Doughty, interview, October 10, 2019.

15. Doughty, interview, October 10, 2019.

16. Aviva Brecher, "Remembering My Mentor, Millie," lecture, Celebrating Our Millie, MIT, Cambridge, MA, November 26, 2017.

17. Aviva Brecher, interview by Madeleine Kline, Margaret MacVicar Memorial AMITA Oral History Project, MIT, Cambridge, MA, September 9, 2017, https://dome.mit.edu/bitstream/handle/1721.3/186178/MC0356_BrecherA_2017.pdf.

18. Doughty, interview, October 10, 2019.

19. Mildred Dresselhaus, interview by Brian Keegan, August 27, 2007, transcript, MIT Infinite History, Cambridge, MA, https://infinitehistory.mit.edu/video/mildred-s-dresselhaus.

20. Nai-Change Yeh, "Mildred S. Dresselhaus (1930–2017): A Fierce Force of Harmony," *PNAS* 114, no. 29 (July 18, 2017): 7478, https://doi.org/10.1073/pnas.1710692114; Mario Hofmann, email to author, November 12, 2019; Helen Zeng, "The Legacy and Impact of Mildred Dresselhaus on Next Generation Career Education," poster, Celebrating Our Millie, MIT, Cambridge, MA, November 26, 2017, https://millie.pubpub.org/pub/9pcicjb1/release/1.

21. Hofmann, email, November 12, 2019.

22. Doughty, interview, October 10, 2019.

23. Mildred Dresselhaus, interview, August 2007.

24. Mario Vecchi, "Millie: A Wave of Happiness," lecture, Celebrating Our Millie, MIT, Cambridge, MA, November 26, 2017.

25. Mildred Dresselhaus, interview by Clarence Williams, *Technology and the Dream: Reflections on the Black Experience at MIT, 1941–1999* (Cambridge, MA: MIT Press, 2003), 360–361.

26. Clarence Williams, email to author, April 11, 2019.

27. Rex Dalton, "Outcry over Scientists' Dismissal," *Nature* 464 (March 11, 2010): 148–149, https://doi.org/10.1038/464148a.

28. Mauricio Terrones, "Mildred Dresselhaus: An Inspiration of Young Generations: A Great Scientist, a Role Model and Colleague," lecture, Celebrating Our Millie, MIT, Cambridge, MA, November 26, 2017.

29. Hofmann, email, November 12, 2019.

30. Doughty, interview, October 10, 2019.

31. Doughty, interview, October 10, 2019. Clara Dresselhaus, email to author, November 19, 2019.

32. Elizabeth Dresselhaus, email to author, November 15, 2019.

33. Clara Dresselhaus, email, November 19, 2019.

34. Obama White House, "President Obama Honors the 2014 Medal of Freedom Recipients," YouTube video, 43:17, November 24, 2014, https://youtu.be/60mECbQJ6TY; "Meryl Streep Biography," Biography. com, A&E Television Networks, April 2, 2014, https://www.biography .com/actor/meryl-streep; "Stevie Wonder Biography," Biography.com, A&E Television Networks, April 2, 2014, https://www.biography.com /musician/stevie-wonder.

35. Obama White House, "2014 Medal of Freedom Recipients."

36. Doughty, interview, October 10, 2019; Laura Doughty, email to author, May 26, 2020.

37. O'Brien, "Pioneering Woman Physicist," 19; Gerald D. Mahan, "Obituary of Peter Clay Eklund," *Physics Today*, September 16, 2009, http://doi.org/DOI:10.1063/PT.4.2105.

38. Doughty, interview, October 10, 2019.

39. Leora Dresselhaus-Marais, interview by author, Cambridge, MA, June 1, 2018.

40. "Vannevar Bush Award," National Science Board, https://www.nsf .gov/nsb/awards/bush.jsp; Doughty, interview, October 10, 2019; Jerome B. Wiesner, "Vannevar Bush: 1890–1974," (Washington DC:

National Academy of Sciences, 1979), 98–100, http://www.nasonline
.org/publications/biographical-memoirs/memoir-pdfs/bush-vannevar
.pdf; "All Members List," MIT Corporation, https://corporation.mit.edu
/membership/all-members-list; "Former Corporation Members," MIT
Corporation, https://corporation.mit.edu/membership/all-members
/former-corporation-members#B; ResearchChannel, "Millie Dresselhaus: In Science, the Real Deal," YouTube video, 51:01, June 15, 2010,
https://youtu.be/qHR51llFZlk.

41. Denis Paiste, "Remembering Arthur R. von Hippel," *MIT News*,
December 12, 2013, http://news.mit.edu/2013/remembering-arthur-r
-von-hippel.

42. Mildred Dresselhaus, curriculum vitae, unpublished.

43. "About the Prize," Kavli Prize, http://kavliprize.org/about.

44. Kavli Prize, "2012 Kavli Prize Laureates in Nanoscience."

45. Clara Dresselhaus, email, November 19, 2019.

46. MIT School of Engineering, "In It for the Long Run," *MIT News*,
July 31, 2013, http://news.mit.edu/2013/mildred-s-dresselhaus-fund.

47. Arlene Alda, *Just Kids from the Bronx: Telling It the Way It Was: An
Oral History* (New York: Holt, 2015), 45.

48. Mildred Dresselhaus, curriculum vitae, unpublished; A. Carvalho et
al., "Phosphorene: From Theory to Applications." *Nature Reviews Materials* 1 (August 31, 2016), 16061. https://doi.org/10.1038/natrevmats
.2016.61.

49. Read Schusky, interview, November 14, 2019.

50. Read Schusky, interview.

51. Leora Dresselhaus-Marais, interview, June 1, 2018.

52. Leora Dresselhaus-Marais, email to author, May 30, 2020.

53. Mildred Dresselhaus, interview, December 29, 2012; Mildred Dresselhaus, interview by Vijaysree Venkatraman, "Reflections of a Woman
Pioneer," *Science*, November 11, 2014, https://www.sciencemag.org
/careers/2014/11/reflections-woman-pioneer.

54. Mildred Dresselhaus, interview, December 29, 2012.

55. Mildred Dresselhaus, interview, December 29, 2012.

56. Clara Dresselhaus, email, November 19, 2019; Elizabeth Dresselhaus, email, November 15, 2019; Leora Dresselhaus-Marais, interview, June 1, 2018; Lauren Clark, "'Rising Stars in EECS' Convene at MIT," *MIT News*, November 14, 2012, http://news.mit.edu/2012/rising-stars-in-eecs-convene-at-mit; Audrey Resutek, "It Takes a Network," *MIT News*, November 18, 2015, http://news.mit.edu/2015/it-takes-network-rising-stars-eecs-1118.

57. Lara O'Reilly, "'What If Female Scientists Were Celebrities?': GE Says It Will Place 20,000 Women in Technical Roles by 2020," *Business Insider*, February 8, 2017, https://www.businessinsider.com/ge-commits-to-placing-20000-women-in-technical-roles-by-2020-2017-2.

58. Leora Dresselhaus-Marais, interview, June 1, 2018.

59. Leora Dresselhaus-Marais, interview, June 1, 2018; Aditi Risbud, "Millie Dresselhaus: Our Science Celebrity," *MRS Bulletin* 42, no. 11 (November 2017): 788, https://doi.org/10.1557/mrs.2017.262.

60. Annie F. Downs, Twitter post, March 5, 2017, 2:02 p.m., https://twitter.com/anniefdowns/status/838464725822410752; Cindy Eckert, Twitter post, February 26, 2017, 10:02 p.m., https://twitter.com/cindypinkceo/status/836048946174767105; Jennifer Granholm, Twitter post, April 29, 2017, 11:56 p.m., https://twitter.com/JenGranholm/status/858530357632798720.

61. Risbud, "Our Science Celebrity," 788.

62. *2020 Diveristy Annual Report* (Boston: GE, 2021), https://www.ge.com/sites/default/files/DiversityReport_02122021.pdf.

63. Julie Grzeda, email to author via GE senior communications director Greg Petsche, February 5, 2020.

64. Bryan Marquard, "Dr. Mildred Dresselhaus, 86, Much-Honored MIT Physicist, Mentor to Female Scientists," *Boston Globe*, February 23, 2017, https://www3.bostonglobe.com/metro/2017/02/23/mildred

-dresselhaus-mit-physicist-and-presidential-medal-freedom-recipient
-dies/FdXiYtUFk6wpgix4n2nz1L/story.html.

65. Marcie Black, "Dr. Millie Dresselhaus: One in Ten Million Scientist, Amazing Violin Player, But So Much More Than That," poster, Celebrating Our Millie, MIT, Cambridge, MA, Nov. 26, 2017, https://millie.pubpub.org/pub/4msu13pj/release/1.

66. "Buckminster Fuller," Mount Auburn Cemetery, https://www.remembermyjourney.com/Search/Cemetery/325/Map?q=buckminster%20fuller&searchCemeteryId=325&birthYear=&deathYear=#deceased=14587502.

67. Sangeeta Bhatia, phone interview by author, December 13, 2019.

68. Leora Dresselhaus-Marais, interview.

69. George W. Crabtree, "Remembering Millie," *MRS Bulletin* 42, no. 6 (June 2017): 464, https://doi.org/10.1557/mrs.2017.130.

70. Brecher, interview, September 9, 2017.

71. Andrew Grant, "Mildred Dresselhaus (1930–2017)," *Physics Today* (February 23, 2017), https://physicstoday.scitation.org/do/10.1063/PT.5.9088/full/.

72. Grant, "Mildred Dresselhaus (1930–2017)."

73. Morinobu Endo, "Prof. Mildred Dresselhaus' Legacy: 35 Years of Collaborating on Carbon Discoveries," lecture, Celebrating Our Millie, MIT, Cambridge, MA, November 26, 2017.

74. David L. Chandler, "A Big New Home for the Ultrasmall," *MIT News*, September 23, 2018, http://news.mit.edu/2018/mit-nano-building-open-0924.

75. Shoshi Dresselhaus-Cooper, email to author, May 16, 2018.

76. Juanxia Wu et al., "Observation of Low-Frequency Combination and Overtone Raman Modes in Misoriented Graphene," *Journal of Physical Chemistry C* 118, no. 7 (January 28, 2014): 3636–3643, https://doi-org.libproxy.mit.edu/10.1021/jp411573c; Materials Research

Society, "Mildred Dresselhaus—Nanocarbons from a Historical Perspective," YouTube video, 31:37, Feburary 14, 2017, https://youtu.be/xHeO9EYJHIs.

77. Materials Research Society, "Nanocarbons from a Historical Perspective," 2017.

78. Jennifer Chu, "Insulator or Superconductor? Physicists Find Graphene Is Both," *MIT News*, March 5, 2018, http://news.mit.edu/2018/graphene-insulator-superconductor-0305; Hamish Johnston, "Discovery of 'Magic-Angle Graphene' That Behaves Like a High-Temperature Superconductor Is Physics World 2018 Breakthrough of the Year," *Physics World*, December 13, 2018, https://physicsworld.com/a/discovery-of-magic-angle-graphene-that-behaves-like-a-high-temperature-superconductor-is-physics-world-2018-breakthrough-of-the-year.

79. Farnaz Niroui, interview by author, Cambridge, MA, May 4, 2018.

80. Julie Fox, "Helping Small Science Make Big Changes," *Slice of MIT*, October 4, 2018, https://alum.mit.edu/slice/helping-small-science-make-big-changes.

81. Mildred Dresselhaus, "New Materials through Science and New Science through Materials," James R. Killian, Jr. Lecture, MIT, Cambridge, MA, April 8, 1987, https://youtu.be/ZOFHDo20YYc.

82. Mildred Dresselhaus, "New Materials through Science."

83. Mildred Dresselhaus, interview by Kelsey Irvin, July 11, 2013, transcript, IEEE History Center, Hoboken, NJ, https://ethw.org/Oral-History:Mildred_Dresselhaus.

84. Mildred Dresselhaus, interview, August 2007.

85. MAKERS, "Mildred Dresselhaus: Queen of Carbon Science," YouTube video, 2:44, April 4, 2017, https://youtu.be/oPfh5nb09yY.

INDEX

Note: Page numbers in *italics* refer to figures or photographs.